Cremona Violins

A Physicist's Quest for
the Secrets of Stradivari

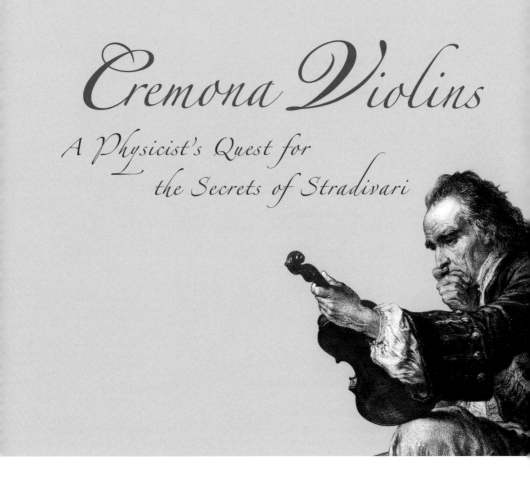

Cremona Violins
A Physicist's Quest for the Secrets of Stradivari

Kameshwar C. Wali

Syracuse University, USA

World Scientific

NEW JERSEY · LONDON · SINGAPORE · BEIJING · SHANGHAI · HONG KONG · TAIPEI · CHENNAI

Published by

World Scientific Publishing Co. Pte. Ltd.

5 Toh Tuck Link, Singapore 596224

USA office: 27 Warren Street, Suite 401-402, Hackensack, NJ 07601

UK office: 57 Shelton Street, Covent Garden, London WC2H 9HE

British Library Cataloguing-in-Publication Data
A catalogue record for this book is available from the British Library.

ISBN-13 978-981-279-109-2
ISBN-10 981-279-109-4
ISBN-13 978-981-279-110-8 (pbk)
ISBN-10 981-279-110-8 (pbk)

Printed by FuIsland Offset Printing (S) Pte Ltd. Singapore

Monona, Achala, Alaka

&

Kashi

Contents

Preface

Cremona violins occupy a unique and storied place in violin history. Andrea Amati and his descendents ushered in an extraordinary period of violin making, which peaked between 1650 and 1750. The most celebrated of the Cremona violin makers, Antonio Stradivari (1644c–1737), brought unsurpassed perfection to the instruments he built, instruments of a completely new quality, leaving behind the legacy of an instrument that possesses all the tonal characteristics of what is recognized today as a "classical Cremona instrument."

For the last three hundred years, well-known luthiers have attempted to replicate the Cremona violins. Although some of them (for instance, the renowned Frenchman, Jean Baptiste Vuillaume) have made excellent copies, the general consensus is that they do not come close to reproducing the distinct voices, carrying power, and responsiveness of the instruments of the old masters. This apparent lack of success has given rise to myths of unknown and unknowable secrets — the source of the wood, its treatment and the particulars of varnish. It has resulted in a vast amount of pseudo-historical and pseudo-scientific literature that is filled with incredible claims.

Likewise, in spite of an immense amount of scientific research along conventional methods, the goal of deriving credible, objectively measurable criteria for the evaluation of the Cremona instruments has remained elusive.

Although studies of the separate components or "mechanical subsystems," of the violin — the bridge, sound post, frequency modes, top and back plate resonances, action of the bow, radiation patterns, wood, varnish, and strings — have provided valuable knowledge regarding how the violins work, they have failed to give any clues about what makes a particular violin stand out among others, let alone the secrets of a Stradivari violin.

William F. "Jack" Fry, Professor Emeritus at the University of Wisconsin-Madison, well-known for his pioneering research in High Energy Physics and Astrophysics, has been pursuing violin research for several decades. He became involved in violin research in the 1960's, having accidentally had the chance to play on a couple of old Cremona violins, one by Allessandro Gagliano and the other by Antonio Stradivari. Amateur violin player that he was, playing occasionally in friendly quartets, for Fry, it was a momentous revelation. *"For the first time in my life,"* Fry says, *"I realized how a good violin can change one's ability to play. The two instruments were so different, and yet so much better than anything I had ever played on."* He became fascinated by the old instruments. How could they be so different from the violins he had played? Was there a scientific explanation for the phenomenon?

This book chronicles Fry's early researches and the gradual evolution in his ideas regarding violin acoustics. Initially, hoping to find some simple answers behind the "secrets" of Stradivari and other Cremona violins, Fry experimented extensively with varnish, frequency analysis, and asymmetries in the graduations of the plates. After numerous experiments, Fry became convinced that the traditional "reductionist" scientific approach was not the answer. Violin acoustics was a complex subject with too many variables. He said to himself, *Stradivari never had any of the modern electronic devices. Think of the great instruments he produced. Why can't we reproduce them with the techniques he used?*

I have known Jack Fry since 1955, when I attended the University of Wisconsin for my graduate studies. Fry had joined the physics faculty a few years earlier and had just founded the High Energy Experimental Group at Wisconsin, which blossomed out as one of the prominent research groups in the world. Fry and his collaborators became well known for their pioneering experiments in K-meson physics, which set the stage for subsequent discoveries of violation of certain fundamental laws in elementary particle interactions.

My own interest in violin research stems from my graduate physics courses in India on acoustics, particularly the work of C.V. Raman on the theory of bowed strings in connection with violin. This work, prior to his discovery of the celebrated Raman Effect that led to a Nobel Prize, had already brought him international recognition. Aesthetically, the violin played in South Indian tradition with its melodic rendering of ragas and its intimate association with the human voice was very dear to my heart. As I became exposed to western music, violins in string quartets and symphonic music stood out for me. During my early years in the United States, when music from India was not easily available, the music of violinists such as David Oistrakh, Jascha Heifetz, and Yehudi Menuhin took over and filled the vacuum.

During my student days and subsequently, Fry and I, have maintained an endearing friendship. The exciting new discoveries in high energy physics had kept us in close contact through our research. I heard stories about his interest in violin research and his quest to uncover the secrets of Cremona violins. I happened to see Fry featured in a PBS sponsored NOVA program titled "The Great Violin Mystery," in which he described his early researches and where he was leading to in his quest for solving the great mystery. Science provided a tool, a way of understanding what the old Cremona masters had in mind when they were making their instruments. It was not beyond imagination that they could have known instinctively and through experiments the underlying physics principles Fry was discovering.

I felt that Fry's work was exciting and largely unknown. I started my own research, interviewing several violin players and violin makers. Over the years, in my conversations with several noted soloists playing in symphonic concerts or in chamber music ensembles, I found invariably their preference for old Cremona instruments, some very well-known, some lesser known. Joshua Bell, Midori and Jennifer Frautschi, all preferred to play instruments by Stradivari or Guarneri del Jesù. They spoke of the superiority of the Cremona violins over the modern instruments in their acoustical attributes. Modern instruments have the necessary power, but lack in dynamic range. They may have one or two good features, but for a player, Italian instruments have a response that makes it easy to play, whereas with modern instruments, one has to work hard to get the desired notes.

Violin makers also spoke of the superiority of the Cremona violins. For them a Stradivari or a Guarneri sound was a mystery. They made excellent copies, but were tradition bound in their instrument making. In the PBS Nova program, Fry described the complex acoustical system of a violin, the principal modes of vibrations and rotations of the plates and how they were affected by minute variations in the thickness graduations of the plates. But violin makers in general seemed unaware of these considerations. My research and preliminary discussions with Fry led me to write an article for the Wisconsin Academy Review.* In the process of writing the article, I journeyed through the fascinating history of Cremona violins, the rise and decline in the art of violin making. I came across the book, *The Violin Hunter* and read the amazing story of Luigi Tarisio, but for whom, "the *Cremona violins created by master craftsmen, geniuses and gifted artists, so much in demand from dukes and emperors for nearly two centuries, would have been extinct.*"

I visited Cremona a couple of times, once with Fry. There was the house on Via Garibaldi, where Stradivari lived from 1667 to 1680 before he moved to a house that overlooked the church of Saint Dominico and the courtyard in front around which the celebrated violin makers lived and worked side by side, building their homes and workshops. The church was destroyed in 1869 and a public garden, renamed the Piazza Roma, now occupies its place and the courtyard surrounded by shops and fast-food restaurants. One of the principal streets is dedicated to Stradivari's memory and a commemorative tablet has been placed on the wall of the home in which Stradivari lived and died. While these sights brought moving memories of what I had read in historical accounts, the highlight of the trip was the visit to the Stradivari museum. Among Stradivari tools, cuttings and wood pieces to be made into violins, Fry found what he calls "scrapers." For Fry it was another momentous discovery, a vindication of one of his most important ideas in his research.

It is a well known fact that the response of the completed instrument has little or no relation to the unassembled pitches of the plates. Whatever the frequencies to which the free plates may be tuned, varnish and gluing together would invariably alter the final response. Hence, for Fry, there had to be a method and a device to control and change the response of the instrument after it was completely assembled. For this purpose, Fry had invented simple homemade "scraping" devices that could be inserted through the f-holes and reach the interior parts of the plates and change the graduations. So Stradivari

---

*William F. Fry: *A Physicist' Quest for the "Secrets" of Stradivari*, Wisconsin Academy Review, Volume 46, Issue 2, Spring 2000.

could have used the same technique! Was it his secret? Fry wondered and was inspired and expressed his full confidence that he would be able to make changes and produce an instrument that had the desired acoustical properties of a Stradivari or a Guarneri del Jesù. That was the moment for me as well to continue my work beyond the article and write this book.

In the first two Chapters of the book, Fry's story is linked to the history, the rise and decline of the art of violin making in Cremona. The third Chapter discusses why the violin is made the way it is made, its components, its structure and its shape, varnish and how it affects the wood and the sound quality of the instrument. It also elucidates the physics of the violin, the mechanism of sound production. The fourth Chapter contains a brief history of violin acoustics research through centuries. The last four Chapters are devoted to Fry's work. From numerous experiments with what Fry calls "old junk violins" from the nineteenth century, which he finds in antique shops and violin stores, he has been able to make predictable changes in the tonal qualities of a given instrument. This is illustrated in the video accompanying the book. The decision to include an accompanying video came from the desire for the reader to hear and experience first hand, as Fry illuminates much of the fine tuning of his process and the predictable changes one can make in the tonal qualities.

Finally, we see in the video, a violin made by Fry, starting from a copy of Stradivari model with a label indicating that it was made in Germany in 1932. In the original form, the measured thicknesses of the central regions of the top and back plates were too thick. It lacked all the specific thickness graduations in the plates that Fry explains (and discussed in detail in the book). The sound we hear (played by Rosemary Harbison) clearly demonstrates Fry's success in duplicating the subtle tonal qualities generally attributed to a Stradivari instrument.

In writing this book, I straddled the difficult task of making Fry's work available to both scientists and to those who might find themselves intrigued on a more general level by the mysteries of the art and science of the Cremona violins. It is my hope that the book will be useful to both groups. However, it has become clear to me this book covers just a tiny part of a subject that is huge in its scope, what it means to reproduce the sound of a Stradivari instrument and other aspects of sociology associated with violin makers and violin dealers. It is my hope this book will serve to keep alive the interest in this fascinating and complex subject.

1

Luigi Tarisio and the Violins of Cremona

What is there about that piece of maple and spruce and ebony that says 'the master fashioned me?'

M. Aldric

It is not too much to say, that with hardly a memorable exception, all the great Cremonese and Brescian fiddles, which now command such prix fous, have passed through the cunning hands of Luigi Tarisio.

Rev. H.R. Haweiss

On a warm spring day in 1827, a man who looked like a tramp, carrying two bags on his shoulders, appeared at the Paris salon of M. Aldric, violin-maker, connoisseur of rare violins and dealer known throughout Europe. The man before him, in tattered clothing and with a bearded face, stood six feet tall with deep blue eyes and a forehead the color of copper. He identified himself as Luigi Tarisio from Milan, Italy; he was thirty-five years old and had walked all the way from Milan. On the way, he had met the violinmaker, Pierre Sylvestre in Lyon, who had directed him to Paris to seek Aldric's salon.

Reluctantly, Aldric let the man enter, who, thereafter, took out from one of his bags a beautiful, small violin without a fingerboard and a tailpiece. Aldric recognized the instrument instantly as the creation of Niccola Amati. Hiding his surprise and joy, he asked the man to show him the other contents in his bag. At this request, the man laid on the counter in front of Aldric — a Maggini, a Francesco Ruggeri, a Storioni, and two Grancinos.

After providing Tarisio with some rest and change of clothes, Aldric assembled friends and colleagues, including George Chanot, Charles Francois Gand and Jean Baptiste Vuillaume — all among Europe's greatest living authorities on violins. For a reported sum of 100,000 francs, together they purchased the masterpieces that Tarisio had brought with him from Italy.

This was the beginning of the rebirth of the Cremona violins, which for all practical purposes had been lost to the world until that point. A very few of them had found their names in the musical registers of the time and almost all of them belonged to one aristocrat or another. In Italy, the instruments, once so famous, lay collecting dust in villas, monasteries and farmhouse attics. In rescuing them from their unfortunate fate, Tarisio had begun a collection that would make him the greatest collector of stringed instruments ever known. On his first trip to Paris, he had brought only a small part of his collection — two sackfuls that he could carry on his shoulders, but it secured him a small fortune and initiated a fabulous career. During the following decades, Tarisio returned to Paris with more and more of the treasured instruments. He became a close friend of Vuillaume and through Vuillaume's patronage, a friend of almost all the well known violin dealers and collectors of the time throughout the continent.

Indeed, Tarisio occupies a special place in the history of Cremona violins. Described in some accounts as a colorful eccentric and a cunning peddler, he was gifted with a sixth sense when it came to tracking old instruments. He was born in Milan in 1792 to a poor family. As a small boy he had lessons in carpentry from his father and lessons in violin from his mother, who had ambitions for him to become a great fiddler in the tradition of Tartini, Corelli, or Pugnani. She inspired him with the contemporary example of Viotti, a blacksmith's son who had become a famous violinist at a very young age. When Tarisio was eight years old, she made him take lessons from a violin teacher in Milan.

However, it turned out that Tarisio was born with a crippled little finger. Within a few lessons, it became evident to his teacher that Tarisio was not destined to be a great violinist. While taking it upon himself to explain the unfortunate situation to his student, he simultaneously planted the seed of a different future for Tarisio. He described to him Cremona, the cradle of the art of violin making. This teacher himself owned a fine Stradivari violin. He spoke to Tarisio about the instrument and about the genius of Stradivari, saying *"There are too few of these beautiful instruments left. Perhaps there lies your destiny — in finding them."*

From his father, Tarisio continued to learn the skills of carpentry and by the age of twelve or thirteen, he became a skilled carpenter, turning out picnic tables and outdoor benches. On his own, he also developed enough skills to fiddle at inns, barns and festivities. Although deprived of a formal education, he learned about the past glory of Cremona and the great works of master violin makers through the stories and rumors he heard. From these tidbits of history, he came to know, for instance, that the art treasures of Italy had been plundered and taken away to France during the Napoleonic wars and that they were much appreciated in that country. Thus he began to dream of the day when he would have the lost treasures of Cremona in his possession in order to take them to France. In 1809, he took the first step towards fulfilling his dream; he visited Cremona.

Figure 1a: Church Saint Sigismondo, a few miles away from Cremona.

Figure 1b: The interior of the church contains the works of almost all the celebrated 16th Century Cremona painters.

He was a young man of seventeen when he walked two days from Milan to Cremona. At the St. Dominico church, he looked around the piazza and the courtyard where, less than a century ago, Antonio Stradivari and other celebrated violin makers of Cremona had built their homes and workshops, and worked side by side. And when he heard about the annual fall festival of the farmers in the parish of Saint Sigismondo, a few miles away from Cremona, Tarisio made his way to the parish to witness the festival, to fiddle and dance and enjoy the Lombardi wines. The festival marked a dramatic ritual in which an artificial bird, Colombina, made of grain and fireworks was set afire and let loose zooming along a wire stretched from the altar in the church all the way to the entrance. The flaming Colombina was believed to be the manifestation of the Lord himself, and as the bird sped along the wire with the golden bundle of grain in its tail sputtering like fireworks, the outstretched hands of humanity struggled to touch the flame. Tarisio shared the excitement of the event, which was followed by the opening of barrels of wine in front of the church.

What came next borders on a miracle. As he drank and fiddled, to his great surprise, he was met by a monk who ushered him into the church and took him to meet a nun named Sister Francesca. Tarisio learned that she was not only the most important nun of the parish but also the granddaughter of Stradivari, the daughter of his youngest son Paolo. As though she knew Tarisio and his destiny, Sister Francesca shared with him her sadness, sadness at the disappearance of the violins of her grandfather and other Cremona masters. She told him how her father and others in the family and priests had failed to persuade the city elders to build a museum commemorating the memory of her grandfather and to preserve and perpetuate the art of violin making. This proposal had fallen on deaf ears. But she was sure that someday the marvelous instruments would reappear; someone would come along and lead a renaissance of the violin. Sister Francesca believed that Tarisio, the young man sitting in front of her, was this person.

If he searched, she told him, he would find abandoned masterpieces everywhere in churches, monasteries, farmhouses, pawnshops and elsewhere. In the Castle Corte Reale in Mantua, the ancestral home of Gonzagas, a prestigious family, which once maintained their own private orchestra, she had heard there were six Stradivaris. And in her father's home, she had seen a violin that defied description in its beauty and perfection, the violin that was certainly her grandfather's most perfect piece of workmanship. She told him how her father, Paolo, disgusted with and angry at the city elders, had parted with it, selling it to a nobleman in Florence, Count Cozio di Salabue. It was in his hands currently and Tarisio must get it back at any cost.

This unexpected encounter and the burden that had been placed upon him left Tarisio in disbelief. Moreover, he was not free immediately to undertake the mission. A commitment had already been made that until he reached the age of twenty-five, Tarisio would be apprenticed to a carpenter in Milan. When he told Sister Francesca about this previous arrangement, she replied, *"You need not worry about the apprenticeship. Here in the church there are ways of settling such affairs."* The next day the monk who had taken him to Sister Francesca the previous day handed him a letter written by the sister to His Holiness, Bishop of Milan. It said,

> *The bearer (of this letter) is Luigi Tarisio, of Milan, who is apprenticed to a carpenter in your diocese. A noble burden has been thrust upon this youth's shoulders; his heart and mind are dedicated to the task of restoring to mankind the treasured creations of my illustrious grandfather, Antonio Stradivari. In our land, torn and bleeding from the wounds of intercine war, sadness and despair have become an intaglio symbolical of all that is evil and bad. The songs and music have vanished from the hearts of our beloved children. The glorious violins and cellos of my grandfather, endowed with an almost heavenly beauty, were fashioned of things ephemeral by a man whose entire years were dedicated, in his own productive manner, to bringing joy and song to the lips and hearts of all mankind; indeed a noble purpose.*
>
> *I myself, Most Reverend Father, have dedicated all my years to the work of Christ. The thread of life has worn bare and my hour is approaching.*

I beseech you to release Tarisio from the bonds of apprenticeship so that he may complete this mission in life.

Along with the letter, he was also handed a violin Sister Francesca had brought with her years ago, one of her grandfather's creations and was told that Sister Francesca had passed away during the night. Saddened and mystified at the strange turn of events in his life, Tarisio returned to Milan with his mind made up to devote his life to finding the lost treasures of Cremona violins. Someday, he fancied and dreamed he would have enough of them to take to Paris and return rich and famous. But upon his arrival in Milan, he found that all this had to wait. Milan was at war, and he himself had received notice of conscription. His ambition thwarted, Tarisio had to leave Milan as a soldier with the army and fight with his regiment against both the French and the Austrians, which he did at Lodi, Piacenza, Casalmaggiori, Guastalla, and Mirandola.

When the war was over, Tarisio was in his mid-twenties. Released from his apprenticeship, his first task was to find a way to earn a livelihood. Carpentry was the only trade he knew. As luck would have it, on his way back to Milan, he encountered a farmer in a village who needed repairs to his house. While doing the job, he found an old violin case containing a violin. Some of the glue had dried out; the fingerboard was loose, with the bridge and the pegs missing; its strings hung on it frayed and useless. But the box was in good condition; it had retained its beauty and had the name of Amati imprinted inside. To Tarisio's delight, the farmer parted with the violin just for the asking. With this treasure in his possession, Tarisio's dream was rekindled. By the time he returned to Milan, six months after the war ended, he had picked up a half dozen or so more of the old violins. In the towns he visited, he came to learn that the contemporary violinmakers had little or no respect for the great works of the past. While this disregard and disrespect for the glorious past did not please him, he was happy in a way for himself. As long as they considered the old violins of little value, his own quest for them would be easy.

After a brief reunion with his parents and family and a period of settling down to carpentry and repairing old violins, he set forth to accomplish his cherished dream. He set off on foot to Mantua, some 120 kilometers away from Milan, and found several old violins the very day he arrived in the town. The city was in ruins, but in the Piazza dell'Erbe, he found a violin shop owned by a person named Dall'Aglio. When Tarisio confided the details of his quest to Dall'Aglio, the shop owner told him to go to Corte Reale, the ancestral home of the Gonzagas, where he would find a caretaker who would give him all the Stradivaris he could carry away. This report confirmed what Tarisio had heard from Sister Francesca several years before on that eventful day. Without wasting any time, Tarisio made it to Corte Reale and found the caretaker priest who guarded the ruins. The priest offered Tarisio food and rest. And then, when Tarisio played a merry tune on the cheap fiddle that he had brought along with him and let the priest play on it as well, the priest disappeared and returned with six wooden boxes, each containing a Stradivari in sad disrepair. In an easy exchange, Tarisio quickly managed to trade a cheap shiny, fiddle for six Strads; the caretaker was content to have a fiddle that he could play on to amuse himself in his lonely hours.

This encounter was the beginning of a pattern — a cheap new fiddle for a priceless old Cremona! He traveled around carrying worthless, modern instruments, which appeared of high-quality and used them as bait for owners of good but often inconspicuous violins badly in need of repair. It took little effort on Tarisio's part to make an exchange. The owners readily parted with what they perceived as junk for a lustrous new fiddle. If parts of an old instrument were to be had, he would buy them for a few coins. *"He bought everything he could lay his hands on,"* says Farga, *"...finger-boards, peg-boxes, bridges, bellies, broken backs, and tail pieces."* Thus, Tarisio became the owner of a formidable collection of old instruments. In the process he became so knowledgeable about the various Italian schools of violin making that *"a single glance sufficed him not only to recognize a Bergonzi, Amati, or Stradivari, but to tell its year of origin as well."*

Of this bounty, Tarisio took only a small portion on his first historic trip to Paris. Although he obtained only a modest amount of money from Aldric and others, it was enough to buy a farmhouse for his parents at Fontanetto near Milan. He could even rent a place for himself, an attic of a second story building, which would serve as his business office and a place to store his Cremonas. Now he could afford a horse and buggy for travel. A few days after his return, he set forth again on his relentless pursuit of the old Cremonas. Luck seemed to favor him all the way. Word was spreading about a young man in search of old violins for new ones. Thus, he learned from a stable owner in Crema that the abbot in the rundown Cathedral had in his possession several old violins. Indeed, the Father Sebastian had several Cremonas and was willing to part with them in exchange for Tarisio's skills as a carpenter, to have some badly needed repairs done in the church. With a day and a night's work, Tarisio was rewarded with two Amatis, a Stradivari, a Storoni and a Guadagnini. Although none of them had strings or any of their mountings, they were all in good condition and had the names of their creators inscribed. In Cremona, he met Carlo Bergonzi 2nd, the grandson of the illustrious Carlo Bergonzi, an apprentice and the last pupil of Stradivari. Bergonzi 2nd was now an old man. He had many great stories to tell to young Tarisio, stories that he had heard from his grandfather about Stradivari's art and craft, his method of choosing wood and varnish. More importantly, he had in his possession several Cremonas that he had used to model his own instruments. He readily parted with them for Tarisio's collection, a collection which eventually became distributed to the rest of the world.

Carlo Bergonzi 2nd also spoke to Tarisio about the last violin that Stradivari had made, an instrument of incredible beauty that had never been played, and which was sold to Count Ignazio Alessandro Cozio de Salabue. Sister Francesca had described the same instrument to him several years before and had charged him to secure it at any cost. Count Cozio was a remarkable man in his own right. Born into the aristocracy on March 14, 1755 in Casale Montferrato in Piedmont, he inherited, when his father died suddenly, a Nicolo Amati violin dated 1688. While still in his teens, Count Cozio developed a great love for the instrument and began to collect old instruments. By 1774, he had at his disposal a superb collection of Cremonas. Out of anger and frustration at the city of Cremona, Paulo Stradivari had sold him all of his father's instruments, as well as his tools and patterns. As a result, Cozio's collection included instruments that would subsequently acquire fame under the names "Le Messie," "The Paganini," "The Viotti." Cozio had also some Guarneris del Jesu, some Ruggeris, some Amatis and some Cappas.

Cozio was not only a collector, but also a man who had a serious mission in his life. He had felt the decline in the art of violin making and determined to rescue it from passing to oblivion. He had heard that Gian Battista Guadagnini, a great violinmaker and son of Stradivari's pupil and assistant, Laurenzo Guadagnini, was living in Turin and at the age of sixty-three was having a hard time making a living. In keeping with his resolution, the young count adopted the old man and commissioned him to make fifty violins. These violins also passed into his collection. And when he died in 1840, having devoted the last decades of his life to the study of Piedmont general history, he left behind an estate that included seventy-four instruments, including the fifty made by Guadagnini. Tarisio, who was certainly aware of the treasures accumulated by the count, succeeded in obtaining most of the treasure for a ridiculously small price and gained possession of some of the gems which he had long coveted, including Le Messie.

Although Tarisio had taken some of the finest specimens of Cremona violins to France over the course of thirty years, he still kept behind in his attic a vast collection of first-rate masterpieces. In talking about his collection, he raved to Vuillaume and Delphin Alard about a special one. *"I have a Strad — so wonderful that one must adore it on one's knees. It has never been played. It's as new as when it left the master's workshop."* For the next twenty years, he would always promise to bring it the following year, and once Alard exclaimed, *"Ah, your violin is like the Messiah — one always waits, and he never appears."* That became the origin of its name when subsequently Vuillaume acquired its possession.

Eventually in January 1855, news came to Vuillaume through a commercial traveler that Tarisio was found dead a few months before in his attic, laying on a sofa, fully dressed, and clasping two violins against his chest. Without a moment's delay, Vuillaume collected all the cash he could lay his hands on and boarded the train to Milan. He found Tarisio's nephews and made them lead him to Tarisio's retreat, the hole in the attic, in Milan. This is how Farga describes what Vuillaume saw and did:

> "*The first violin which Vuillaume took from its case was a magnificent Antonio Stradivari from the master's best period. Two more instruments, in perfect condition, had been made by J.B. Guadagnini. The fourth one was an especially beautiful Guarneri del Gesu, whose varnish seemed to sparkle in the gloom of the room. Then came a light-brown Carlo Bergonzi with golden reflexes, famous as the master's best piece of work. The greatest surprise was the sixth box, the opening of which took some time. When Vuillaume at last took out the violin and held it up in the poor light of the candles, he cried with joy. There it was, the Messiah violin — safe and sound as though Stradivari had just completed it, a shimmering jewel, mysteriously alluring with the pent-up magic of its tone.*"

In the end, Vuillaume found no less than 144 violins including two dozen Stradivaris, violas, and cellos in the dirty attic. For a sum of 80,000 French Francs, he bought the whole treasure! Through Vuillaume, the illustrious, highly skilled luthier in his own right, Tarisio's vast treasure of Cremona found its way into the hands of dealers, connoisseurs and collectors of violins, and through them into the hands of great violinists. Indeed, it is profoundly ironic that the Cremona violins created by master craftsmen, geniuses and gifted artists, so much in demand from dukes and emperors for nearly two centuries, would have been extinct if it had not been for Luigi Tarisio, a man of humble birth — with only the skills of a carpenter, from the small village of Fontaneto, near Novora in Piedmont, Italy.

The Rise and Fall of the Cremonese Art of Violin Making Antonio Stradivari (1644c–1737)

That indescribably sweet voice, voice of an angel and yet ringing with the dear familiar sound of the earth, with earthly passions, joys and woes and ecstasies.

Paul Stoeving

More than any other instrument, the violin has attracted the attention of historians, musicologists, novelists, poets, operas, films and scientists — not excluding the physicists. Colorful descriptions of its form and beauty are abundant in literature. To quote Stoeving further:

> *Was ever a form more perfect symbol of the tone, the body of the soul within? A delicious play of curves and colors — the noble sphinx-like head from which it rolls down or unfolds itself... that amber color deepening to a rich, an almost reddish brown towards the center where the sound-life pulsates strongest, quickest! A corner of a Titian canvas, is it? Yes, or Rembrandt's. And behold the fine fiber of the wood shining through the varnish like delicate roses through my finger nails! What can be finer??*

These poetic descriptions contrast with those in scientific literature. For instance, Carleen Maley Hutchins, one of the pioneers in violin research, describes the violin essentially as,

> *a set of strings mounted on a wooden box containing a volume of almost closed air space. Some energy from the vibrations induced by drawing a bow across strings (precious little energy, it turns out) is communicated to the box and the air space, in which are set up corresponding vibrations. These in turn set the air between the instrument and the listener into vibration; in other words, they produce the sound waves that reach his ears.*

This, of course, sounds very prosaic, but physics has its own poetry and finds violin a marvel of complex acoustics that embodies a great deal of simple physics, more of which will be discussed in the latter part of this book.

In spite of a great deal of research, speculation and controversy, the origin of the violin is shrouded in mystery. It is amusing to note, for instance, that its origin is traced to India, to King Ravana from the epic Ramayana, the ten headed demon, who was also apparently a great singer and musician and invented the instrument called "Ravanastron," the first instrument played with a bow. But modern accounts agree that violin, more or less in its present form, appeared predominantly in Italy, in the 16th century. Two schools flourished, one in Brescia, founded by Gasparo di Bertolotte (1540–1609) (known as Gasparo da Salo) and the other in Cremona, founded by Andrea Amati (1505c–1577).

Gasparo da Salo is believed to be one of the earliest violin makers. His pupil, Giovanni Paolo Maggini (1580c–1630), is well-known for the fine instruments he made towards the end of his career. The Cremona school, however, dominated the scene for the next two centuries with instruments of a completely new quality, leaving behind the legacy of an instrument that possesses all the tonal characteristics of what is recognized today as a "classical Cremona instrument." According to Isaac Vigdorchik, this came about because of the discovery of a new pitch relationship between the top and bottom plates. Brescian makers tuned their plates to a half tone relationship, whereas the Cremona violin makers used a differential of a whole tone. Furthermore, succeeding generations of Cremona violin makers introduced improvements and modifications in the acoustical treatment, external design, arching, and air capacity of the body. Through experimentation, they also established the appropriate thickness graduations in particular areas of the top and back plates. "*The magnificent instruments that these makers created in the 17th century have served as an inspiration to generations of craftsmen, down to the present time. It is no surprise that that these instruments were prized and appreciated during the lifetimes of the makers.*" says Vigdorchik.

Cremona, on the banks of river Po, is visible from kilometers away because of the tower of the cathedral, apparently the highest church tower in Italy. Contemporaneous with the Brescian school, in 1539 Andrea Amati established his workshop in the heart of Cremona, in the square dominated by the church of St. Domenico. Amati's instruments became well known in the musical world of the time; he had supplied an entire concerto of string instruments to the Parisian court of Catherine de'Medici. His son, Gerolamo (1548–1630), following his father's tradition, consolidated the family's fame with instruments of extraordinary beauty and sweetness of sound that came to be known as "the Great Amatis." But it was his son, Niccolo Amati (1596–1684), who ushered Cremona into its golden period of violin making. His instruments excelled in both the aesthetic and acoustical qualities over those of his grandfather.

Niccolo lost his parents and two sisters in 1628–30 when two successive famines followed by the plague left him alone, depressed and despondent. Gradually, when he started to work again around 1640, he faced an over-whelming demand for his instruments. He desperately needed help and having no children of his own, he was forced to break the family tradition and begin employing non-family members and provide training for them. Thus he entered a new phase in his life and became the teacher of almost every important violinmaker of the late 17th century that included Franceso Ruggeri (1645–1700), Andrea Guarneri (1624c–1698) and the most famous of all, Antonio Stradivari (1644c–1737). Generally conceded to be the greatest among the violinmakers, Stradivari brought unsurpassed perfection to the instruments he built. All in all, the period from 1650 to 1750 was the most illustrious period in the history of violin making, the heyday of all the celebrated violin makers. They lived and worked side by side, building their homes and workshops around a courtyard in front of the St. Domenico Church (Fig. 1).

During this golden period of approximately 150 years, Italy was under Spanish rule, and Italian history of this period is mainly the history of its music, a period in which its music dominated the western world. It was the age of the madrigalists, the birth of opera and its development from the entertainment of princes into a flourishing commercial industry. And the history of Italian music of the 17th century is inseparably bound with that of the great Italian violin makers. As we read in *The Violin Makers of the Guarneri Family (1626–1762)*, "*Supremacy of the violin was connected in a peculiarly intimate way with the artistic character of the music itself. The 17th century is the century of Baroque architecture, and the violin is a typically Baroque instrument. Painting and sculpture show us that throughout the ages musical instruments exhibit the architectural lines of their ages no less than articles of domestic furniture. The shape of the violin, the curves of its outline, and the convexities of its back and top*

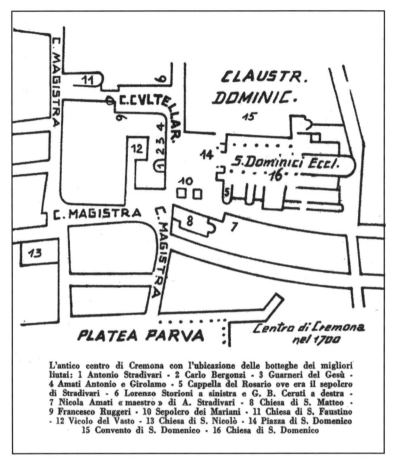

L'antico centro di Cremona con l'ubicazione delle botteghe dei migliori liutai: 1 Antonio Stradivari - 2 Carlo Bergonzi - 3 Guarneri del Gesù - 4 Amati Antonio e Girolamo - 5 Cappella del Rosario ove era il sepolcro di Stradivari - 6 Lorenzo Storioni a sinistra e G. B. Ceruti a destra - 7 Nicola Amati «maestro» di A. Stradivari - 8 Chiesa di S. Matteo - 9 Francesco Ruggeri - 10 Sepolcro dei Mariani - 11 Chiesa di S. Faustino - 12 Vicolo del Vasto - 13 Chiesa di S. Nicolò - 14 Piazza di S. Domenico 15 Convento di S. Domenico - 16 Chiesa di S. Domenico

Figure 1: The court yard in front of the St. Domenico church, where the celebrated violin makers lived and worked side by side. (From *La Cremona Di Stradivari* (*Stradivari Itineraries*), *Editrice Turris-Cremona —1988*)

are characteristically Baroque — so much so, indeed, that the layman might well imagine that they were dictated more by artistic than by acoustical reasons."

Stradivari's family origins and the place and date of his birth are not known. It is conjectured that he was probably born in some outlying parish near Cremona, where his parents might have fled during the disastrous plague that swept through Lombardy in 1630. The year of his birth, 1644, is deduced by dating the age of his later violins. The first certain date in Stradivari's life is the

25th of June 1667, when his marriage deeds to Sinorina Francesca Tiraboschi was published in Saint Agatha's church. The marriage took place on the 4th July 1667 and the couple started living at 55 Corso Garibaldi, not far from the church of St. Domenico. A stone marker on this building states that the Stradivaris lived there between 1667 and 1680.

Figure 2: Stradivari's house in Piazza Domenico (now Piazza Roma) is no longer standing but is captured here in a nineteenth-century engraving. (From W.H. Hill *et al.*, *Antonio Stradivari*, His Life and Work (1644–1737) 1963.)

The house had three floors and a garret, the ground floor of which served as the workshop and the rest as residential quarters. Six of Stradivari's sons were born in this place, before he moved his shop and pupils in 1681 into a house that overlooked the church (Fig. 3). It had a similar layout but was more substantial with spacious cellars and an attic between the third floor and the garret. He lived and worked the rest of his life in this house. Francesca died on May 25, 1698, and a year later, he married Antonia Maria Zambelli, who bore him five more children.

Figure 3: The picture of the house as at present at Piazza Roma. (From W.H. Hill *et al.*, *Antonio Stradivari*, His Life and Work (1644–1737) 1963.)

It is generally believed that Stradivari was apprenticed in Nicolo Amati's workshop and learned his skills of violin making from the great master. As incontrovertible evidence, the earliest violin made by Stradivari dates to 1666 with the label bearing the words, *"Antonius Stradivarius Cremonensis Alumnus Nicolaii Amati, Faciebat Anno 1666"* (Made by Antonio Stradivari of Cremona, pupil of Nicolò Amati.). The very next Stradivari violin that has survived is the one made in 1667; its label, however, reads simply, *"Antonius Stradivarius Cremonensis Faciebat Anno 1667."* Thereafter, there is no mention of Amati on his labels. These early violins show amazing skill in woodcarving, the purfling inserted with such perfection that one is lead to believe it required some special skill and expertise outside the art of violin making. As further evidence, the decorated instruments he started making in the late 1670s display beautiful artistry. They carry beautifully inlaid purfling, elegant tracing around their sides and on their scrolls.

This has led to the conjecture that Stradivari was trained as a woodworker before he became a violin maker. Indeed, the house he lived in from 1667 to 1680 was owned by Francesco Pescaroli, a woodcarver and an inlayer by profession. It is very likely that Stradivari learned his skills in wood carving and inlaying from Pescaroli before he went to Amati's workshop, most probably first as a decorator for Amati's instruments. There are indeed one or two ornamented instruments made by Amati in the mid 1650s that bear distinctive resemblance to the later violins made by Stradivari.

During the extraordinary career that spanned over seven decades, it is estimated that Stradivari made some 1,116 stringed instruments, of which 700 Stradivari violins survive. However, the exact number is impossible to ascertain since the sales of many Strads are conducted privately between a few discreet dealers and collectors. In *Antonio Stradivari*, W. Henley records four periods in Stradivari's violin making: 1) *The Amati Period, 1666–1690* during which, following initially Amati's style, he begins his own independent approach, striving for *"greater beauty of form and the higher pinnacle of tone."* 2) *A Time of experiments, 1690–circa 1700.* He continues making gradual changes, working the scroll in more boldly and in greater detail. He modifies his pattern, narrowing its shape, making the back plate a little longer by one-eighth of an inch to 14.25 inches. He experiments with different color varnishes. 3) *The Golden Period 1700–circa 1720*. During this period, he brings together his decades of experimentation, combining *"the powerful tone of the Brescian instruments with the clear sweet sound of the Cremonese. His instruments now had wider outlines, faultlessly spread out arching joining together from all directions forming a beautiful shape. All materials very carefully selected. Scrolls less deeply cut and beautifully carved. Fine golden-yellow varnish of matchless quality, sometimes with a further coat of bright red varnish. Inside work impeccable. Linings were made of willow. The purfling was set at a distance of 4 mm from the edge. From the time onwards there were about 200 violins, 10 violas and 20 'cellos known."* 4) *Period of Decline 1720–1737*. Although outstanding in tone, instruments during this epoch lack the perfection of the earlier period. Some instruments are probably made by others under his supervision as indicated by the inscription *"sotto la disciplina 'D'Antonia Stradivari.'"* The varnish of the instruments of this period is generally brown and less transparent than that applied to his previous instruments.

With the successive deaths of Stradivari in 1737, Guarneri 'del Gesu in 1744, and Carlo Bergonzi in 1747, the glorious century of violin making came to an end. Within a period of approximately three decades, Cremona ceased to exist as the center of violin making. The events of war, the defeat of Spain and the conflicts between European powers saw Cremona occupied by various French and German armies. It entered a period of unrest and fear of Austrian domination and the possibility of being drawn into Austria's conflicts with Prussia and Turkey. By 1773, violin making in Cremona became seriously affected by the projects of reforms of Maria Teresa, Empress of Austria. The reforms hit the mercantile world hard, including those of arts and trades, and upset the economy of the whole region around Cremona. Italians loved music, loved violins, but they had started acquiring cheap imports from Germany that satisfied their needs. The instruments once so famous and prestigious were no longer in demand and had practically disappeared. Antonio Stradivari was all but forgotten.

However, his imposing three-story home, where he lived and worked for over fifty years and created the fabulous instruments, stood in the Piazza St. Domenico across from the church (Fig. 3). In 1729, eight years before his death, he had purchased the tomb that belonged to the descendents of a noble Cremonese family situated in the chapel of the Rosary of the church. It became the burial place for his wife and himself and other relatives in subsequent years.

After the death of his brothers, Francesco and Omobono, Paulo, the youngest son of Stradivari became the heir of the Stradivari fortune. A cloth merchant by profession, he was intensely devoted to the memory of his father and wanted to convert the Stradivari house and his workshop into a museum. *"He offered to endow the museum,"* Silverman writes, *"almost eighty of his father's instruments, as well as with other historic relics: an old family bible reputed to contain, on the flyleaf, the exact formula of Stradivari's varnish; tools, patterns, and documents, many of which bore the names of kings, emperors, dukes and princes, and even a pop."* When his pleadings, as well as those of his brother Giuseppe, fell on the deaf ears of the mayor and city councilmen, Paulo in utter despair and anger decided to part with his treasure to Count Cozio di Salabue. He had already sold ten of his father's finest instruments to him. Now he offered to sell all the patterns, tools and measures in his possession to the Count on the condition that they must be removed from Cremona. After some negotiation, the sale took place and Cremona lost some of its finest treasures. Left behind were still a few violins including one which was to be hailed eventually as the masterpiece, an instrument so perfect that it acquired the name Le Messie (The Messiah). Finally, on his death bed in 1776, he instructed one of his sons to sell all those remaining violins to the Count and see that the last treasures were taken out of Cremona.

The sad story did not end there. The bodily remains of Stradivari and some of his descendents had stayed intact till the year 1869 in the chapel of the church. For almost a hundred years there was the talk of the demolition of the church whose walls had kept some masterpieces of the history of Cremona from 1184. The 23 chapels and the altars of the church had been frescoed and stuccoed by some of the famous artists over the centuries. But over the years it had fallen into a state of decay (Fig. 4). Finally in 1869, the town authorities decided that it had reached a stage that was positively dangerous and it was time to pull it down along with the adjoining deserted convent and create a public garden on their sites. The demolition began in the spring. We have an extremely moving account of the demolition process by an eye-witness, Signor Mandelli:

In the summer of the year 1869 the work of demolition of the fine church of St. Domenico was making progress; in fact, the great apses of the church, the tower, and the Chapel of Christ had already disappeared under the never-ceasing blows of the pickaxe, the dull sound of which was slowly re-echoing along the pillars and against the vaulted ceilings of the aisles of the chapels which still stood untouched. It was a sacrifice imposed by modern requirements, and the imperious exigencies of civilization and hygiene. When the work was once in full swing, the masons cared not which part was to be attacked; the

Figure 4: A photo showing the last glimpse of the church of S. Domenico. (From W.H. Hill *et al.*, *Antonio Stradivari, His Life and Work (1644–1737)* 1963.)

pickaxe, incessantly in use, had already rained down its blows upon the Chapel of the Rosary, demolishing the cupola by Malosso and the ceiling by Cattapane. I still remember it, and vividly recall the remorseless destruction of this work of a past splendor.

Signor Mandelli continues on to say that he was present on the day when the demolition reached the tombs. Several distinguished citizenry had assembled around the tomb of Stradivari. He heard the following words from one of the gentlemen present. *"There is such a confusion of bones, without any special mark whatever, that it seems useless indeed to make any further search."* Mandelli heard the name of Stradivari several times, but he was too young then, he says, to grasp the significance of the search the gentlemen were making. During the succeeding days he saw men with baskets clear the tomb of all the human bones found within it — skulls, tibias, thigh bones, and ribs. No attempt was made to sort them out into individual skeletons. The workmen themselves interred the bones outside the city with the exception of a few skulls that were saved as souvenirs by the men in charge of overseeing the demolition, Signors Ferdinando Rossi and Francesco Ferrari.

Thus the tomb of Stradivari disappeared, but the name-stone did survive, and it is now preserved in the Municipal museum. In place of the church whose walls depicted the history of Cremona in masterpieces frescoed and painted by some of Italy's most famous artists, one now finds a public garden, renamed the Piazza Roma and surrounded by shops and fast-food restaurants. We should be thankful that one of the principal streets is dedicated to Stradivari's memory and a commemorative tablet has been placed on the wall of the home in which Stradivari lived and died, bearing the inscription,

HERE STOOD THE HOUSE
IN WHICH
ANTONIO STRADIVARI
BROUGHT THE VIOLIN TO ITS HIGHEST PERFECTION
AND LEFT TO CREMONA
AN IMPERISHABLE NAME AS MASTER OF HIS CRAFT.

A B

1687 Amati period "Kubelík"

This 1687 Stradivari violin was given to the Czech soloist Jan Kubelík by a wealthy English patroness. It was used by Kubelík until he acquired the Emperor Stradivari violin of 1715. Sold by the University of California Berkeley in the May 2003 auction to benefit their music department, it fetched $949,500, which at the time was a world record for a pre-1700 period Stradivari violin.

– Tarisio, fine instruments and bows' website

A   B

1714 Golden period The "Soil"

This violin was acquired by Itzak Perlman in 1986 from Lord Yehudi Menuhin. Its original neck is preserved in the Stradivari Museum in Cremona. It was recorded by Hill in 1902 as in the ownership of Mr Soil, and as "a specimen of the highest order". "The "Soil" has an enormous range of tone color and its vast, warm, cavernous sound is familiar to all concert goers. "It is in the opinion of many informed listeners the greatest sounding Stradivari of them all" writes Charles Beare.

–Capolavori di Antonio Stradivari, Cremona, 1987

A
B

1736 later period The "Muntz"

Pathetically portraying the veteran's work, as the Hill brothers described it in 1902, this instrument made one year before he died has corners and edges of very square appearance. Nevertheless, as Charles Beare describes it, this instrument has "plenty to admire" and "a first class reputation for tone". This was an instrument kept by Stradivari's younger son Paolo, who sold it to Count Cozio di Salabue in 1775.

–Capolavori di Antonio Stradivari, Cremona, 1987

The Anatomy of a Violin and the Mechanism of Sound Production

The violin is a marvel of science —
mathematics, physics, chemistry,
acoustics — and also the miracle of
a passion, the love of music. It is that
rare mixture, the synthesis of emotion
and intellect, of passion and science.

Joseph Wechsberg

Violin is an enormously complex instrument. As the music is played, the box has to respond to the rapidly changing frequencies, and respond almost instantaneously. Yet, as Carleen Hutchins describes, the violin, in essence, is an instrument with a set of strings mounted on a wooden box containing a volume of almost closed air space. When a violinist draws the bow across the strings, he or she sets them into vibrations that are communicated to the box and the airspace by which corresponding vibrations are set up. These in turn generate the sound waves that reach the listener. From such simplified descriptions it is hard to conjure up the complexities of the structural and other details, and also the fine art of fashioning these instruments.

The design of the box, as one might imagine, could be spherical, cylindrical, square or even triangular in shape. In reality, however, the violin's shape suggests a far more artistic than scientific origin in its form. It seems to have taken on the likeness of a human female torso — long neck, rounded shoulders, narrow waist and supporting pelvis. Other intricate details, like the deliberately upturned corners and the graceful holes on the top plate only add to this perception of physical elegance.

Indeed for a violinmaker, the design and construction of a violin involves a multitude of variables. No formula, however detailed, can describe the all-important quality of each element, each piece of wood that comes together to make a violin. A violinmaker would say that violin making is a matter of experience and intuition, artistry and craftsmanship, and that there is no such thing as an "exact copy" of a famous model, if for no other reason than the fact that identical pieces of wood cannot be found. Even so, a scientific reasoning with the help of physics, does provide a deeper understanding for most of the design features. Consciously or otherwise, through experimentation or happy accident, the great Cremona violinmakers must have been aware of these scientific reasons in making their violins.

The basic components of a violin are as shown in the accompanying figure (Fig. 1). We note the two arched plates of the resonator box, the *top* and the *back*. They are held together by the *ribs*, which follow the contours of the plates. The top has two holes in the shape of the letter script F and hence they

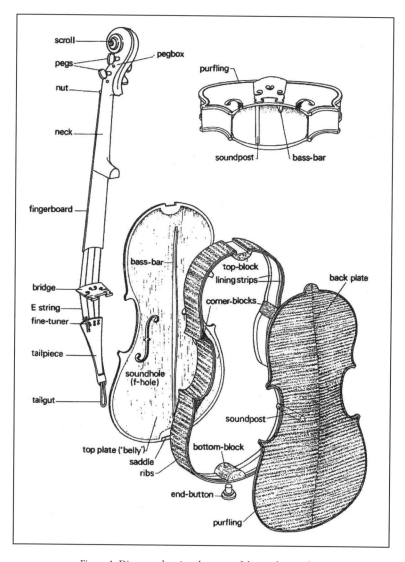

Figure 1: Diagram showing the parts of the modern violin.

are called, the *f-holes*. It also has a wooden piece attached along its length, called the ***bassbar***. There is a pencil shaped wooden piece, called the ***soundpost*** that fits between the two plates. We also see the unfretted fingerboard and the highly arched ***bridge*** over which the four strings of the violin are stretched. The fingerboard has the *tailpiece* and the neck at the two ends; the neck is attached to the *peg box* and ends in a *scroll*. The corner and the **end blocks**, the lining strips and the **purfling** complete the description of the anatomy of a violin.

In several conversations with William F. "Jack" Fry, we discussed the physics of the violin; we considered why the violin is made the way it is, its components, its structure and its shape. We discussed a simplified model to understand the basics of the highly complex mechanism of violin sound production and the critical role that the various components play in this process. What follows includes a summary of these discussions.

I. The Instrument

(i) The Structure of the Box and its Components

We will begin by taking a closer look at the box, the plates and their shapes, and the kind of woods of which they are made. For the top plate, the wood used could be soft spruce, pine or fir. The rest of the instrument, the back plate, ribs, neck and scroll, are all made of hard maple. There is, of course, a good reason for the choice of the particular kinds of wood to make the various components. The sound that any instrument produces, like any other acoustical device, is related to the way in which various components of the instrument move mechanically. In the case of the violin, the sound radiated by the instrument is principally due to the vibrations and the motions of the two plates, the top plate and the bottom plate.

As we will see later, a simplified model can be constructed to understand the sound producing mechanism in a violin. This model explains the movements within the instrument in terms of three principal mechanical modes (or the way in which the plates move) in three different frequency ranges. The back plate of the violin plays an important role in producing lower frequencies. At higher frequencies, it remains stationary, supporting the sound post so that it can act as a fulcrum around which the top vibrates and radiates the higher frequencies. Therefore, it makes sense that the back plate is made from the relatively denser maple and the top plate from the lighter spruce woods. The stiffness of the plates is another important parameter, which varies from wood to wood, but any wood is stiffer along its grain than in the transverse direction.

When we consider the amount of force the plates experience due to the high tension in the strings, stability emerges as an extremely important factor. To achieve this, the back plate is first cut from one board, which is then sawed in the middle. One half-piece is turned around and rotated through one hundred and eighty degrees. The two pieces are then glued together (Fig. 2). This way, if the board tends to warp in some dimension (as is the natural tendency of wood), the two parts that are cut from the single board and turned around, will want to warp in opposite directions. Consequently, the dimension in which the wood tends to warp remains unchanged, rendering greater mechanical stability to the plate and the instrument (Fig. 3). The top also usually consists of two parts, each 4.5 inches in width since finding one piece of high quality wood of sufficient width can be difficult.

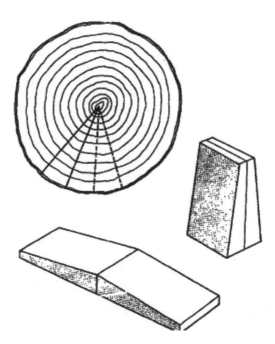

Figure 2: Radial section split (cut on the quarter) and glued base to base after turning one half-piece around through one hundred eighty degree.

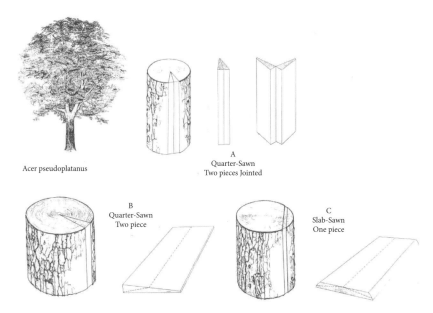

Figure 3: Making the back plate of the violin from maple. Two methods, one from two pieces, and the other from one piece.

(ii) Arching of the Plates

The arching of the plates is crucial in making violins, as it affects the sound and the tone of the violin in many subtle ways. It is also an essential aspect of the beauty of the violin's shape. Stradivari and other violinmakers of Cremona experimented with arching and used it to change the tone and quality of the sound. To produce the desired arching, one begins with the shape of the plates shown in Fig. 2. By gradually carving the outside, the arched top is created (Fig. 4). Finally, the inside is carved to bring the plate to the desired shape. This is the way one produces the arching with desired variations in the thickness and curvature, leaving a convex shape outside and a concave shape inside the plate. However, it is worth noting that the curvature decreases steadily to either side. The edges are nearly flat. The plate is arched in two dimensions of the plane, along the length and also along the perpendicular. Arching along the length makes it stiff for flexure in the direction perpendicular to the curvature in one dimension of the plane. Arching in the perpendicular direction makes

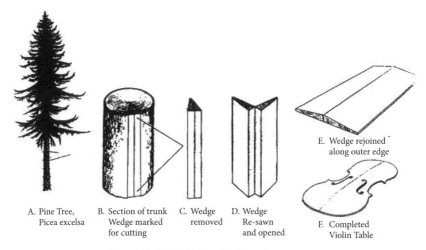

A. Pine Tree,   B. Section of trunk   C. Wedge   D. Wedge
   Picea excelsa   Wedge marked   removed   Re-sawn
                      for cutting   and opened

E. Wedge rejoined
   along outer edge

F. Completed
   Violin Table

Figure 4: Making the top plate from pine or spruce.

it stiff for flexure perpendicular to curvature in the other direction. The central part of the plate is the stiffest. It has the shorter radius of curvature, whereas the radius of curvature increases as one goes toward the edges.

Arching produces many significant effects essential to the strength and sound of the violin. Without arching, the thin top plate, additionally weakened by the f-holes, would not be able to support the pressure of the strings transmitted through the bridge. But most importantly, arching increases the volume or the loudness of the sound of the instrument. If a flat plate were pushed at the center, the deflection would be limited to a very small region around the center, causing only a small amount of displacement of air inside the box. If, on the other hand, the plate was arched and the central part was stiff, most of the deflection would occur along the edges. The pressure is spread over a larger area. Since a greater area of the arched plate is involved in displacing the air inside the box, the plate then acts like a piston displacing a considerably larger mass of air. So, arching is a way to increase stiffness of the plates without increasing thickness and adding mass.

Highly arched instruments, therefore, can be lighter. They radiate higher frequencies more easily, and their sound will be more brilliant and pure. Yet they generally lack in depth compared with the later, more flattened instruments. It is interesting that the earlier instruments of Amati and Strainer were more highly arched than the later instruments of Stradivari and Guarneri. Stradivari, early in his life, influenced by Amati, his teacher, initially made highly arched instruments. With the passage of time, presumably with increased understanding of the acoustics and probably the realization that arching effects can be compensated by thickness variations, he made flatter, less arched instruments. It is also true in general of Guarneri that he reduced the arching.

(iii) The Ribs

The "ribs" that close the box take on its shape, and therefore, they are curved. The curvature strengthens the wood enormously for flexures. Simply because they are bent, the ribs are extremely strong and rigid, and they are strongest where the radius of curvature is the smallest. That is, where they are most bent. Thus, it follows that the weakest regions will be those where they are bent the least. The ribs are made from several thin pieces of wood, about 1.5 mm in thickness. They must necessarily be thin in order to bend them along all the corners. To fit the desired contour, they are first wet and then placed on a hot iron, where they are bent into the required shape. Acoustically, one can think of the ribs as essentially rigid. Their flexibility is not much of a factor in the motion of the top and the back plates. However, the curved winding shape of the box is not simply a matter of aesthetics or artistic fancy. It is one of the most critical factors in the way the plates move, and it also makes the box structurally stable.

(iv) The Bassbar and the Soundpost

Two more components of the box, the *bassbar* and the *soundpost* are crucial to the mechanics of sound production in a violin. The *bassbar* is rigid and stiff and is glued to the top plate underneath the left foot of the bridge. The soundpost, 2/3 the size of a pencil, is wedged perpendicularly between the top and the back plates. It is almost, but not quite, underneath the right foot of the bridge. It is displaced toward the player by about two millimeters. As we see later, its position is extremely critical.

(v) The Scroll and the Peg Box

Superficially the *Scroll* and the *Peg Box* may look more like decorative pieces than components that play any crucial role. The strings stretch between the peg box at the end of the neck and the tailpiece. Actually the bridge is the end point of the motion of the strings at the tailpiece end. However, one must allow some motion there since the bridge has to rock to transmit the force to the box. At the far end, we have — in addition to the static force due to the tension in the string — a great deal of oscillatory force, due to the oscillatory motion of the strings. This motion has to be eliminated or minimized; otherwise the effective acoustical lengths of the strings would be greater than the terminating lengths.

In order to minimize this motion, the neck has to be stiff. Since it is long, there is a limit to how stiff the neck can be made, even taking into account the fact that the fingerboard's hard ebony does add some stiffness. One way to accomplish this is to attach a big hunk of mass that has large inertial impedance to hold the end points of the strings totally fixed. The *Scroll* is that big hunk of mass. It plays a much more crucial role than what one might think. If one detaches the scroll, the violin does not at all sound the same.

The *Peg Box* itself adds some mass, but not as much as the *Scroll*. Its role is critical in terms of its mechanical properties in preventing the pegs from slipping and giving way under the tension of the string.

(vi) Protruding Wings and Purfling

The *Protruding Wings* also serve an important purpose. Notice how the corner blocks fit in exactly where the wings are. Now, as noted before, the inside of the violin box is like the inside of a guitar. The lining that is put on to strengthen the extremely thin ribs gives a smooth outline inside. As it is bent around, the lining has essentially zero curvature and is, therefore, weakest at the corners. The corner blocks that fit into the wings strengthen those exact spots. So the wings and the corner blocks work together to strengthen the box mechanically.

The striking delicate line that follows the outline of the edge of the violin following the shape of the box is called **Purfling**. This **Purfling** consists of three thin strips glued together with the central black strip made of hard wood (like ebony) and thinner layers of maple (or occasionally willow) on both sides. The whole strip is 1/2 to 1/3 mm thick and is pressed in and glued in a narrow groove cut in the plate. It is extremely difficult to stretch because the grain in the wood runs along the length of the strip. Since it is inlaid in the plate at both upper and lower parts of the plate, the strip strengthens the plate in the dimension perpendicular to the grain of the wood in the plate. As a result, **Purfling** decreases the possibility that the plate will crack along the grain as the wood dries out or in the event of mechanical stress.

(vii) The Shape of the Box

There is a great deal of controversy concerning how the shape evolved. Many believe that the viol belongs to an older family of instruments than the violin. The viol family does not have corner blocks and protruding wings, but the viols are narrowed in the middle. It makes sense that the narrowing in the middle probably came about because of the need to allow space for the bow as it is played over the strings, for otherwise the bow would hit the edge of the instrument. This seems like an obvious reason. However, the narrowing also has a very strong effect on the way in which the plate moves because it is precisely at this place on the instrument that the f-holes are cut. Since the f-holes weaken the plate, narrowing strengthens it.

II. The Mechanism of Sound Production

The study of the mechanics of sound production in a violin involves an understanding of how violin vibrations are generated and distributed and how the vibration of violin plates leads to sound radiation. Basically, as the violin is played, the strings are displaced to the right or left by the action of the bow. The vibrating motion of the strings is then communicated to the bridge, which in turn transmits it to the feet of the bridge. The feet of the bridge then communicate the vibrations to the box.

If we think how fast the strings are moving during actual playing conditions, it seems almost impossible to analyze the motion of the box and the plates in terms of a finite number of modes or types of motion. But according to Fry, as far as the mechanics of sound production is concerned, one can separate the motion into three principal modes, the *breathing mode*, the *rocking mode* and the *tweeter mode*. These three principal modes provide a great deal of insight into the underlying physics and the various physical factors that come into play.

(i) The Breathing Mode

The *breathing mode* is the basic mechanism needed to understand the action of the box when the sound is in the low frequency region involving frequencies and fundamentals ranging from 200 to 1000 Hertz. Bowing sets the string into vibrations horizontally perpendicular to its length. Now, suppose the string is pulled to the right. This exerts a force on the bridge making the left-foot go up and the right-foot go down. The bassbar that is under the left foot of the bridge transmits that force to the bulk of the top, which then moves up. At the same time, the right foot pushing down transmits a force through the sound post to the back plate, which therefore moves down. To put it simply, the top moves up, the back moves down. Obviously, the opposite happens when the string is pulled to the left. The two plates are moving in opposite modes. The top moves down, the back moves up.

When the string is displaced to the right, the bridge rocks to the right; the box expands and the air flows in. When the string is displaced to the left, the box collapses and the air flows out. The motion of the air in and out of the f-holes follows the mechanical motion of the string. The box is alternately expanding and contracting, following the motion of the string to the right and the left respectively. When the box is expanding, it takes in the air through the f-holes, and when it is contracting, it forces the air out through the same f-holes, as with inhaling and exhaling when we breathe. The f-holes, therefore, act as the nose and the mouth of the violin; this is what Fry calls the *breathing mode*.

Furthermore, the velocity of the air coming in and out of the f-holes depends on both the displacements of the plates, as well as the ratio of the areas of the plates compared with the areas of the f-holes. Since the area of the plates is much larger than that of the f-holes, a small displacement of the plates will cause a big displacement of air through the f-holes. The result is that the velocity of the air coming out of the f-holes is much bigger than the velocity of the plates themselves. And since the energy varies as the square of the velocity, there is an enormous amplification of the energy going in and out of the f-hole. In this way, the breathing mode compensates for the fact that the area of the violin plates alone would be a very inefficient radiator for these low frequencies.

(ii) The Rocking Mode

As frequency rises, in the mid-frequency range (2000–5000 Hz), the inertial property of the back plate becomes significant enough to minimize its vibrations. It holds the sound post stationary and prevents it from moving up and down. Now, as noted earlier, the sound post is not directly underneath the right foot of the bridge but displaced from it. And since it is stationary, it acts as a fulcrum to allow a teeter-totter motion of the top plate when it rocks as a consequence of the movements of the bridge up and down. Fry calls this the *rocking mode*. In this mode, the rotational properties of the bassbar play a significant role. It influences the ability of the top plate to rock, and this impacts the frequencies in the range of a few thousand Hertz, which is the mid-range voicing of the violin.

(iii) The Tweeter Mode

Once frequency increases into the high frequency range (>5000 Hz), the moment of inertia of the bassbar becomes too high for the bar to rotate or rock. Hence, the rocking motion of the top also comes to a stop. At this point, the area located around the right foot of the bridge comes into play. This area drives the motion of a resonant system located on the right side of the right f-hole near the ribs, what Fry calls a *high frequency tweeter*, essentially the source from which most of the high frequency components are radiated in the frequency range of ~10,000 Hertz. At these high frequencies, the wavelengths are extremely short, in the centimeter range. And the shorter the wavelength, the less surface area is needed to radiate sound of the corresponding frequencies. To be more precise, the effective radiating surface area goes inversely as the square of the wavelength. Consequently, only a small area of the top plate is sufficient to radiate the sound in this *tweeter mode*.

Thus, the three principal modes — the *breathing*, the *rocking* and *tweeter* — provide a good physical basis for understanding the mechanism of sound production in a violin. However, it must be emphasized that the actual mechanism is much more complex, and the description in terms of the above three principal modes is not wholly realistic, as it involves several simplifying assumptions. Consider, for instance, *the breathing mode*. The model assumes that all the air, due to the change in the volume of the box, goes through the f-hole. Now, the ratio of the velocity of the air that goes through the f-hole to that of the air due to the motion of the plates is inversely proportional to the two areas. That is to say, the smaller the area of the f-holes, the greater the velocity of air coming out of the f-holes. And, since the energy radiated is proportional to the square of the velocity, one arrives at the absurd conclusion that the energy output can be made arbitrarily large by reducing the area of the f-holes. Careful analysis shows that after reaching a maximum, the energy radiated begins to go down as the area of the f-hole is reduced. Indeed, there is an optimum area of the f-holes that would radiate maximum energy.

The idea that energy output can be made arbitrarily large assumes that the forces are always strong enough to drive all the air out of the f-holes. However, the forces generated by the string vibrations have a limit, and the frictional turbulent properties of the air going in and out of the f-holes produce enough impedance (resistance) to stop the velocity from increasing beyond a certain limit. In fact, one can determine an optimal area for the f-hole by matching the impedances between the string, the plates and the resistance of the f-holes. If the area is bigger or smaller than this optimal area, the volume of the sound goes down. The peak of the sound occurs at the correct optimal area. To the credit of the great violinmakers, they discovered this through experimentation a long time ago. Evidence for this experimentation can be seen in the museum in Cremona, where there are some strange violins that are not "normal looking" at all. These instruments have a variety of shapes and sizes for the f-holes.

Furthermore, the idealized model implies that the energy of the air coming out of the f-holes is unaffected by the energy of the air moving at the surface of the top. Imagine that the air coming out of the f-holes has enough time to go around the box in such a way that the energy gets distributed evenly, leaving no differentials in pressure to generate the sound waves. In such an instance, the overall size of the box becomes a crucial factor. For the ***breathing mode*** to work, the box must have a physical size such that the time required for the air to move around the box is considerably larger than the period of oscillation of the air moving in and out of the f-holes. To understand this point, consider the model of a tin can with a flexible diaphragm on the top containing a little hole. When the top is pushed, the air comes out, and when it is let out, the air goes in. If this motion is slow enough for the air to move around the entire can (that is, if the diaphragm is held pushed long enough and let out slowly) the differential pressures are zero everywhere, and no sound radiated. On the other hand, if the motion of the air in and out is fast enough, the differential pressures near the diaphragm will generate sound. In a similar manner, in order for the breathing mode to work efficiently, one needs an optimal box size combined with an optimal size for the f-hole.

Yet another important physical aspect of the ***breathing mode*** is worth noting. The volume changes of the box driven in this mode make it act like a Helmholtz resonator, an almost closed vessel with a protruding spout that Helmholtz used to recognize the pitch of a musical note. The essential principle behind such a resonator can be understood by considering the example of a COKE bottle. It is a matter of common experience that if one blows across the open end of a COKE bottle, it gives rise to a resonant whistle of a definite pitch or frequency. A physical model corresponding to the COKE bottle could be envisioned in terms of a simple harmonic oscillator consisting of a mass attached to a spring. The air in the bottle, as it compresses and expands, acts as a spring and the smaller quantity of the air in the throat as it is pushed up and down acts as the mass. Given that the frequency of a simple harmonic oscillator depends upon the square root of the ratio of the spring constant to the mass, we can estimate the frequency of the resonant pitch of the bottle by knowing the volume of the air in bottle and the volume of air in the throat.

Similarly, in the case of the violin, it is the air in the box that corresponds to the spring, and its compressibility determines the spring constant. The air coming in and out of the f-holes determines the mass. Increasing the volume of the box increases the amount of air it contains; this decreases the compressibility (the spring constant) and, hence, the frequency. Reducing the volume creates the inverse effect, and the violin sounds more "soprano". If the f-hole opening is narrowed (enlarged), the mass decreases (increases) and, hence, the frequency increases (decreases). The natural resonance frequency for most violins lies generally somewhere around the note C on the G-string and can be controlled, since it plays an extremely important role in the low frequency range of a violin. This resonance should not be a prominent resonance that drowns out the notes around it. The width of the resonance depends upon energy losses. In this process, the dominant energy loss is due to the viscous forces of the air moving in and out of the f-holes. As the f-hole opening is narrowed and elongated, the air moves faster. This increases the energy loss due to viscous forces, making the resonance broader. On the other hand, the greater energy loss results in the decrease of the intensity of the output. Thus, a delicate balance has to be maintained among various factors. The shape of the f-hole is extremely critical, and the shape is one of the reasons the tonal qualities of the Guarneri instruments are different from those of the Stradivari instruments. Although the shapes of the f-holes look alike to the naked eye, there are enough differences to shift the tonal qualities, lending each instrument its distinctiveness.

The shape and the placement of the f-hole near the right foot of the bridge raises a whole set of other problems. Its long and thin shape cuts along the fibers. As noted earlier, woods in general have large asymmetries in their stiffness. They are 15–20 times stronger along their grain, that is, along their fibers. This property of wood provides its primary strength and support.

Consequently, it is critical that the placement of the f-hole be such that it does not cut the fibers under the foot of the bridge and those over the soundpost (the soundpost fibers). If the f-hole touches those fibers and cuts them, the violin is essentially destroyed. In contrast, the left foot of the bridge sits on the bassbar, which is long and covers almost the whole length of the plate. The fibers in the area around the left foot of the bridge are strengthened, and the force is distributed and transmitted to most of the area of the top plate.

(iv) Varnish and its Role

A great deal of discussion in the literature surrounds the varnish and the process of its application. Some violinmakers and researchers even claim that varnish is the sole secret of the old masters. According to Fry, most of this is misleading. However, there is no doubt, that if you were to play a violin when it is unvarnished, and then play it after it is varnished, the sound would be quite different. There is no mystery about it. The change concerns two factors. Varnishes both add mass and change the stiffness of the woods as they harden. The stiffness of the varnish is not much different than that of the wood along the stiffer dimension, that is, along the grain. However, most varnishes are stiffer than the wood across the grain. So, the primary effect of the varnish is the change in stiffness it brings about in the transverse direction. This reduces the intrinsic asymmetry in stiffness characteristic of wood. The effect of the varnish depends upon the type of varnish, the control of its hardening conditions (these depend upon temperature and other factors) and, most importantly, the control of the depth of its penetration into the wood. All these factors have a noticeable and significant effect on the sound of the violin.

4

Some Historical Notes on Violin Research over Centuries

The sensation of a musical tone is due to a rapid periodic motion of the sonorous body; the sensation of a noise to non-periodic motions.

Hermann Helmholtz

Antiquity to the 16th Century: Pythagoras and Relative Length Ratios

The recognition of consonances and dissonances in musical experience goes back to antiquity, long before anything was known about *pitch numbers* and their relation to frequencies of periodic motions. The musical consonances described as the octave, the fifth and the fourth that resulted in a pleasant sensual experience were known long before Pythagoras (ca 570–497 B.C.). But it was Pythagoras, during the sixth century B.C., who established through his *monochord* that such consonances were characterized by the ratios of simple whole numbers.

The monochord can be thought of as a single string kept under tension by attaching weights to one end, the other end held fixed. A movable bridge could be used to alter the length of the string that was set in motion by striking or plucking (Fig. 1). Pythagoras discovered that for consonant musical tones, the lengths had to be in the ratio of some simple whole numbers. Thus, the octave corresponded to the ratio of 2:1, the fifth to the ratio 3:2, and the fourth to the ratio 4:3. He also noted that the shorter the length, higher the pitch, establishing the law that pitch was inversely proportional to length.

Pythagoras's discovery and the reduction of the subjective auditory experience of consonances to ratios of simple numbers had a profound effect beyond musical history. It laid the foundation of mathematical physics as a bridge between physical experience and numerical relations. Subsequently, this idea dominated Pythagoras and his followers, who saw such simple numbers and their ratios pervading all natural phenomena. Celestial bodies, stars, planets, the sun and the moon, created sounds in their motions just as the *earthly bodies* in motion. The music of the celestial sounds, depending on their distances and speeds, had the same concordant harmonious relations as the monochord. The idea that a circular motion of the stars and planets was harmonious led to the "music of the spheres" that governed the large-scale organization of the universe (Fig. 2).

The relative length ratios that corresponded to consonant musical intervals in monochord experiments formed the basis of musical theories from antiquity

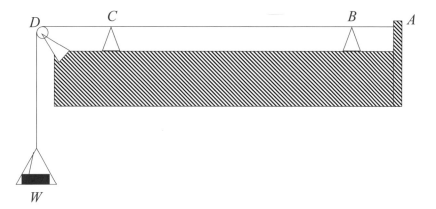

Figure 1: Monochord.

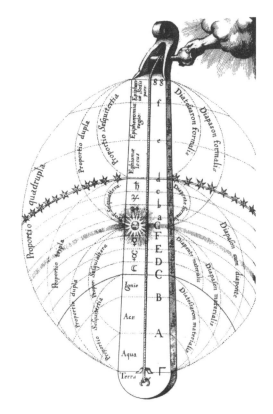

Figure 2: Music of the spheres.

past the middle ages to the 16th century. Up to this period, the theoretical discussion of harmony concentrated itself solely on the consonances produced by the mathematical ratios of the various string lengths and the division of the octave. Soon, the ratios of simple numbers were extended to other parameters of vibrating strings. Thus, the ratio 2:1 of string lengths to tune an octave was also supposed to be valid for tensions of strings of equal length. Testing theoretical ideas with observations and measurements was yet to come.

The 17th to the 18th Century: The Galileis, Mersenne, Chladni and Vibration Frequencies

After the 16th century, subsequent discoveries in the physics of sound based themselves on the crucial realization that the sensation of pitch corresponded to frequency, and it is precisely the frequency of vibration or any other periodic motion that lends itself to a rigorous quantification. Furthermore, it became clear that consonances or dissonances had nothing to do with the aesthetics of simple ratios of lengths. As Galileo Galilei would explain in *Dialogues Concerning Two New Sciences*:

> *I assert that the ratio of a musical interval is not immediately determined by the length, size, or tension of the strings but rather by the ratio of their frequencies, that is, by the number of pulses of air waves which strike the tympanum of the ear, causing it to vibrate with the same frequency. This fact established, we may possibly explain why certain pairs of notes, differing in pitch produce a pleasing sensation, others a less pleasant effect, and still others a disagreeable sensation.*
>
> *Such an explanation would be tantamount to an explanation of the more or less perfect consonances and of dissonances. The unpleasant sensation produced by the latter arises, I think, from the discordant vibrations of two different tones which strike the ear out of time. Especially harsh is the dissonance between notes whose frequencies are incommensurable.*

Galileo Galilei

This simple yet profound realization led musicians and musical theoreticians to reconsider modes and musical scales and their consonances. It also led to more careful experiments with vibrating strings. Thus, Vincenzo Galilei (1520–1591), a renowned composer and lutenist who also happened to be the father of Galileo Galilei, first started experimenting with strings of different materials, different thicknesses (different cross-sectional areas) and different weights (different tensions). Father and son made measurements and discovered that frequencies could depend on parameters other than the length. They noted that the tone of a string may be sharpened (that is, frequency increased) in three different ways, namely, by shortening the length, by stretching it (increase the tension by increasing the weights) and by making it thinner. If the tension and thickness of the string remains constant, one obtains the octave by shortening the length to one-half. If length and thickness remain constant, doubling the stretching weight to obtain the octave is insufficient; it must be quadrupled. If the length and tension remain constant, the thickness of the string must be reduced to 1/4 of its size to produce the fundamental.

Similar considerations held for other intervals. To obtain the fifth, the ratio of the lengths would be 3/2, but the required tension in the ratio of 9/4, thinning by a factor of 4/9 from the fundamental.

During the same period, Marin Mersenne (1588–1648), a French priest, mathematician and philosopher, was also working on the subject of pitch and frequency. Like Galilei, he recognized that the string transmitted an equivalent number of vibrations in one second to the bodies of instruments and to the air in motion. Referred to as one of the "fathers of acoustics," he published his observations and his experiments in his 1636 treatise *Harmonie Universelle*. In this groundbreaking work, he described the pulsations of two strings nearly in tune and determined the first frequency of an audible tone ever given at 84 vibrations per second. He also provided a precise mathematical formulation of the law of vibrating strings by stating, *"Under idealized conditions frequency is inversely proportional to length and directly proportional to the square root of tension and to the square root of the cross-sectional area."* The law thus generalized the pitch-frequency relation of a vibrating string to any frequency not

Mersenne

merely restricted to a few consonances. He further separated the consonance/dissonance question into separate fields: physics (vibrations and partials) and effects (use of consonant chords and of dissonant ornaments).

Another experiment in *Harmonie Universelle* worth noting consisted of finding notes from precisely cut samples of wood of the same size but from different species of trees. He noted,

> *-It would be too difficult to probe all the different kind of bodies present in nature to find out how their tones and sounds differ, and that is why I will only mention the ones I have tried myself, starting with fir, sycamore, service tree, willow, hornbeam, oak. Alder, walnut tree, Chinese wood, ebony, beech and prune wood, to which anyone can add as many as he wants. Now, fir sounds a diminished fourth higher than sycamore, wild cherry a minor third above sycamore, and service tree a minor third below sycamore. Willow is at unison with wild cherry. Hornbeam is a tone higher than sycamore, the same as oak. (…) Chinese wood is at unison with fir, but it has a much clearer and more resonant sound, such as being nearly the same as metal.*

This experiment indicates that Mersenne was aware of the common practice of "tap tones" used in instrument making. We do not know if his research had any impact on instrument makers, but his monumental contributions certainly set the stage for theoretical and experimental research of musical instruments and opened the field to a new science.

Indeed, Ernst Florence Friedrick Chladni (1756–1824), a physicist and an amateur musician, inspired by the developments in natural sciences, sought their application to music. He studied and wrote about the longitudinal vibrations of strings and rods and traveled giving lectures with practical demonstration of his experiments with the vibration modes of plates. His lectures created widespread interest in acoustical phenomena and led to experiments by others with the shape of the violin and its modifications. Chladni's most important contribution was his development of the Chladni technique that he used to observe modes of vibrations of metal and glass plates of the glass harmonica-like instruments that he himself had designed. Later, Felix Savart used Chladni's technique to study the normal modes of vibrations of violin plates.

The 19th Century: Savart and Helmholtz; Savart's Experiments with Violins

The first scientist known to have conducted direct experimental studies of violins is the Frenchman, Felix Savart (1791–1841), a 19th century Parisian

physician and a physicist. His father, a manufacturer of precision instruments for physicists, encouraged Savart's work in acoustics and the study of vibration theory.

Savart's association with the violin craftsman Jean-Baptiste Vuillaume (1798–1875) greatly furthered Savart's research. At a time when violin makers were taking the tops off to install heavier bassbars, Vuillaume, one of the great violin craftsman and a great intellectual,

Felix Savart

was among those who made such modifications on many fine early Cremona violins. As noted in Chapter 1, Vuillaume was also a shrewd businessman and had acquired a good number of Cremona violins through Tarisio. He shared Savart's zeal for scientific research and readily collaborated with him, placing some of the best violins including those of Stradivari, Guarneri and others at Savart's disposal.

Savart's main research focused on the sounds in the free top and bottom plates. As a physicist, he knew that two plates of the same size but of different types of wood or with different characteristics would emit sounds of different frequencies when vibrating. To determine the frequencies, he would fasten the plates in a special vise and run a bow along the lower edges of the plates. He invented a special device to determine the frequencies of the sounds emitted.

Using the Chladni technique, Savart also identified the normal modes of the disassembled plates. The Chladni technique consisted of mounting the plate horizontally, sprinkling it with a fine powder, and making it vibrate by bowing with a rosined violin bow at various points along its edges. In response to the vibrations, stationary wave patterns were set up on the plates. At specific frequencies of resonance, a distinct pattern of nodal (not vibrating) regions and anti-nodal (vibrating) regions of the plate emerged. He found that the observed pattern of "plate resonances" or normal modes was unique to each plate. The pattern depended on the plate's physical properties of stiffness and mass distribution. Experiments conducted on several Italian instruments established that the air in violin boxes of Stradivari and del Gesù emitted the pitch 512 cps. He also discovered that the basic pitch of the top plate, lying between C-sharp and D, was a whole tone lower than the back plate with a pitch lying between D and D-sharp. Among his other scientific discoveries, Savart established that sound traveled 15 to 16.5 times faster in spruce than in air. In maple it was only 10 to 12 times faster.

Savart is also famous for disregarding the traditional violin shapes and using his knowledge of physics and acoustics to build violins with optimal tonal qualities. Under his directions, violins in the trapeze shape were built, which had no arching, straight f-holes, and a straight bassbar in the center of the top plate. (Fig. 3) When the tonal qualities of such a violin were demonstrated in the presence of scientists and artists from the Academy des Sciences and the Academy des Beaux-Arts, it stood well in comparison with a classical violin. It displayed great purity and evenness of tone and was heralded as a remarkable achievement and a great scientific advance. However, Savart's violins failed to become widely accepted. For violinists as well as lovers of violin music, the trapeze-shaped violins lacked the aesthetic appeal that seemed to matter as much as the tonal qualities. Another reason may have been that it would have appeared ridiculous for a violinist to play a trapezoidal shaped violin in a concert hall.

Yet, Savart's use of modern scientific concepts and experimental methods had an impact on the art of violin making. François Chanot, for instance, thought that the corners of the standard violin inhibited the vibrations, and therefore, designed a guitarlike violin, modifying the shape of the f-holes into C-holes and placing them close to the sides. One of Chanot's violins was tested by the members of the Academy des Sciences and judged superior to a Strad in its tonal qualities. However, these fine tonal qualities did not last long. Deteriorations set in for reasons not completely known.

Figure 3: Savart's trapezoidal violin; Chanot's violin.

Helmholtz and the Helmholtz Resonators

Savart's experimental scientific approach was further extended by Hermann von Helmholtz (1821–1894), a physician, physicist, and philosopher well-known for his seminal contributions in diverse fields of physics. He was

concerned with not only the mechanics of vibrations and resonances but also with the effects of audible tones on the listener. His monumental work, *On the Sensations of Tone* (1877) encompassed the interfaces between the physiology of sensation, psychology, and physics of perception.

Helmholtz invented ingenious devices known as *Helmholtz resonators* (Fig. 4) to observe and analyze the composition

Hermann von Helmholtz

Figure 4: Helmholtz resonators.

of a complex tone in a variety of instruments, including those of the violin. They consisted of spherically shaped bodies made of glass or metal containing a volume of air, each with two openings; one, a hole through which sound could enter and set the volume of air into vibrations, the other, a conical or funnel shaped end that could be fitted into the ear canal with the aid of warm wax. Each resonator, by varying its size, could be tuned to a specific frequency so that the observer listening through the resonator, could hear clearly any partial component frequency of the complex tone that might coincide with the frequency of a given resonator.

Helmholtz was intrigued by the fact that the area of the ear through which sound is transmitted is small compared with the wavelengths of sound waves that generate audible musical notes. It is so small that it has no way to sense the variations in degrees of density and velocities that are occurring in the propagation. Nonetheless, the ear is capable of distinguishing musical tones arising from different sources. Helmholtz wondered about the means that the ear employs and the special characteristics of the musical tones relevant for this happening. He thought that the relative strengths and other characteristics of the higher partials (overtones) distinguish the quality of a complex tone. To study the properties of a complex tone, he thought of generating such a complex tone by the superposition of several pure tones. He invented an apparatus, the *Helmholtz resonator* that could simultaneously set into vibrations a set of tuning forks tuned to different frequencies. He could control their intensities and their superposition timings by clever switching devices (Fig. 5). By using the Helmholtz resonator, he found that he could produce a complex tone that was independent of the *phases* of the component pure tones, depending upon only the amplitudes and frequencies of its components. This was an important discovery in the perception of the quality of sound.

Helmholtz is also noted for his studies of the forms of vibration of individual points on the bowed string with the aid of a vibration microscope, an idea first proposed by the French physicist Lissajous (1822–1880). He established the

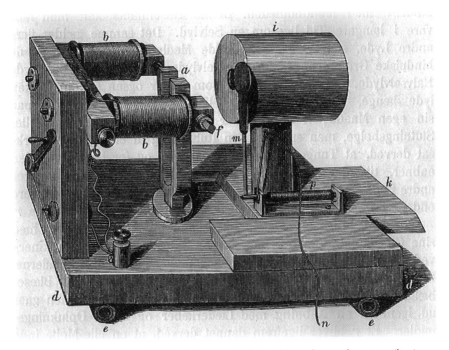

Figure 5: Helmholtz apparatus, the *Helmholtz resonator* that could simultaneously set into vibrations a set of tuning forks tuned to different frequencies.

sawtooth waveform characteristic of the bowed string. As the bow is drawn across the string, a rather sharp "corner" or "kink" shuttles back and forth around what appears to be a lens shaped trajectory (Fig. 6a). As the kink passes the bow in one direction, it dislodges the string from the bow and reverses the motion of the string. The opposite happens when the kink passes from the reverse direction. It catches the string and moves it along. This slip and stick motion gives rise to a discontinuous velocity wave form (Fig. 6b) with one episode of slipping and one of sticking in each cycle. Helmholtz was the first to recognize that the periodic impulse produced by the slip-stick action of the rosined bow hairs on the rosined string sets up a regime of oscillations in which the upper vibration components are maintained in simple harmonic relation to the fundamental.

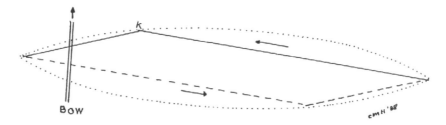

Figure 6a: The saw tooth wave form characteristic of the bowed string. The dotted line shows the lens-shaped path of the kink that shuttles back and forth.

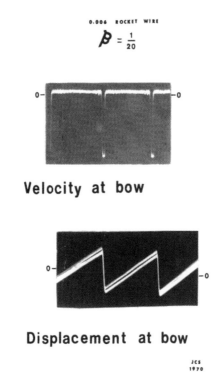

Figure 6b: The discontinuous velocity wave form due to the slip and stick motion. Oscillograms of the actual motion of the bowed string.

The 20th Century: Helmholtz and Beyond

Helmholtz's wide-ranging work influenced Lord Rayleigh (John William Strutt, 1842–1919) to develop Helmholtz's theory further. Rayleigh believed

Lord Rayleigh

that additional experimental data was needed to clarify the details of the theory. His theoretical work included not only the violin string but also the vibrations of membranes, plates, shells and bells. It brought together new and earlier works and laid the foundation for the enormous expansion of the knowledge of acoustics that was to follow in the twentieth century. He was also one of the early users of analogies in electrical circuit theory and forced vibrations of acoustical resonators and other systems, concepts that proved very useful in the pioneering work of John C. Schelling (1892–1979).

Instrument maker Russian Anatoly Leman (1859–1913) stands out as one of the most important researchers in violin acoustics of the late 19th and early 20th centuries. Leman traveled to Italy and other countries in Western Europe to study the works of great masters and was reputed to have analyzed more than 4500 instruments to establish a scientific method for making an "ideal" instrument. He recorded his observations in numerous brochures and books, the most significant being *The Acoustics of the Violin,* published in 1903. In pursuit of scientifically based principles for the acoustical construction of bowed instruments, *"he proposed for the first time the signal importance of determining not only the pitches of the top and back plates but also those of several distinctly expressed and localized areas."*

An outstanding and skillful instrument maker, Leman made some 200 violins, cellos and violas known for their *"graceful patterns and their mellow, but penetrating sound,"* says Isaac Vigdorchik. *"They also reveal his skill as a craftsman, his excellent taste and a strongly expressed individuality. Although Leman's instruments do not stand up to those of the greatest makers of the past, they do compare favorably in many respects with those of the lesser Italian makers of the 18th century. Occasionally one finds examples which are so fine in tone quality one can almost believe that he had mastered the art of Italian violin making."*

With the advent of twentieth century technology and related theoretical advances in optics, photography and electronics, violin research, like other fields, veered towards "reductionism." A study of the separate components, or "mechanical subsystems," of the violin — the bridge, soundpost, frequency modes, top and back plate resonances, action of the bow, radiation patterns, varnish and strings — has dominated research. It is beyond the scope of this book to give a full account of the vast research except to provide a few highlights to give a flavor of this kind of research.

A. The Bowed String Theory and Experiments

C. V. Raman

Influenced by the works of Helmholtz and Rayleigh, Chandrasekhara Venkata Raman (1888–1970) studied the bowed strings of violin and cello in great detail and provided the first comprehensive theoretical discussion of the dynamics of bowed strings. He built an automatic bowing device, which could simulate the bowing techniques of a violinist quite closely. With the aid of this device, he measured the effects of bow speed, bow force, and the effect of the distance of the bow from the bridge on the modes of string vibration. With the idea of developing a dynamical theory, he varied these physical parameters and found the laws governing their relationship to each other. For instance, he found that the minimum bowing pressure varied directly as the speed of the bow and inversely as the square of the distance of the bow from the bridge. With the help of photographic records, he brought to light different types of bowed string vibrations and proposed a kinematical model to describe a hierarchy of periodic motions including the basic "Helmholtz motion." He also constructed a dynamical model to explain his observations, assuming the string to be an ideal, flexible string which terminated in real, frequency independent resistances and a nonlinear frictional force.

Raman is also credited with explaining the so-called "wolf-note," or "wolf-tone," a phenomenon extremely disturbing for a violinist. According to Raman, the phenomenon arose, when the fundamental of a note matches a strong resonance in the instrument body. The steady, stable Helmholtz motion corresponding to the bowed note is interfered with due to the energy loss from the string to building up the instrument resonance. Thus, a warbling or stuttering noise — a wolf-note — is produced, consisting of an alteration between periods of Helmholtz stick-slip motion and periods in which there are two or more slips per cycle. The effect is more obvious when the note is being played near the minimum bow force.

Since Raman, the theory of bowed strings has been studied in great detail. The idealized Helmholtz theory of sharp cornered (seesaw) waves is theoretically unstable and unlikely to occur. More sophisticated theories have been developed incorporating the "smoothing out" of the corner, taking into account the finite width of the bow and the dissipation of energy. But analytical methods and numerical solutions can go only so far. Various schemes of computational modeling of the bowed string have been developed and have grown to become an active field of research.

B. Tap Tones, Plate Tuning and Beyond

Common knowledge tells us that the violin makers of the past used "Tap Tones" as a guide in the construction of their instruments. That was probably what led Savart to wonder about the tap tones of the plates, the sounds they should have, before they were assembled. He then proceeded to test the sounds of free plates and use Chladni patterns to determine the normal modes. After measuring the vibrations in dozens of plates, Savart arrived at their main tap tones and found that the top and the back differed by a tone to semitone, with the top frequency being higher than the back. Needless to say, with the advent

Carleen M. Hutchins

of new technologies, the Chladni pattern technique has been developed further and used to map the normal modes of free plates and their dependence on thickness and other constraints. Consequently, the free (unassembled) top and back plates can be trimmed and tuned to the desired normal modes. Violin makers have used the techniques and the findings of Savart and Chladni in building their own instruments with some success. In particular the work of Carleen M. Hutchins has shown that certain modes — the first, the second and the fifth — are important in free plate tuning. She has pioneered the use of free plate tuning in making her own instruments and The New Violin Family, as well as popularizing the method through her research.

The Chladni pattern technique itself underwent a modern rejuvenation when loud-speakers with powerful electronics capable of inducing sufficient plate motion became readily available. To observe the normal modes, a free plate is placed horizontally over a loud speaker with the inside of the face facing upwards. A sine-wave signal is directed through the speaker, sweeping across the frequency range of interest and causing powder sprinkled in the plate to assume characteristic patterns at discrete frequencies that are unique to the plates. As Bissinger describes, *"This meant that the plates did not have to be clamped for mechanical excitation, but could float over the speaker on small foam supports and be acoustically driven, giving a much better approximation to "free-free.""* The advent of lasers combined with holographic interferometry techniques have further helped to provide precision data concerning the normal modes. This has led to the subject of *modal analysis,* the theoretical understanding of the normal modes as the eigenmodes and eigenfrequencies of the plates.

C. Modes of the Completed Violin Body

The knowledge of certain characteristic relations in the eigenmodes and eigenfrequencies of free plates makes it possible for an individual violin maker to consistently fashion his own instruments. However, there is no obvious way of knowing what happens to these modes when the plate pairs are assembled into the completed complex vibratory system and how they are correlated with the tonal characteristics of the final system.

In an effort towards a more complete understanding, Savart attempted to use Chladni pattern techniques to determine the normal modes of the arched convex outside surfaces of the plates. He found that the technique would not work since the powder simply rolled off. This may have led Savart to experiment with flat outside surfaces, as was the case with his trapezoidal violins.

In the 1920s, with the advent of electronic devices such as the oscilloscope, the heterodyne analyzer, and various recording devices, it became possible to map the normal modes of the arched top and back plates, and the first researcher to do this was Herman Backus in the 1930s. In an effort to study the modes of the completed violin, Backus's work was continued and further developed by his student Herman F. Meinel. The latter's research enabled the verification of theories proposed at the time concerning violin acoustics and led to proposals about how to improve the tonal quality of a given instrument in a prescribed frequency range. However, it became clear that the problems involved in making predictable changes were too challenging. The tonal qualities of an instrument depended upon its constructional details, the modes of vibration, relative stiffness, damping, and the choice of varnish. These factors varied from instrument to instrument. Not surprisingly, Meinel found that a small change that markedly improved one instrument would adversely affect another.

Eventually, technical advances based on hologram interferometry were developed in the late 1960s and made it possible to obtain more detailed and comprehensive information concerning the vibration modes of the whole violin. It also became possible to obtain visual images of the modes of vibration leading to experimental analysis and mathematical modeling with greater and greater sophistication.

The first experimental *modal analysis* of a violin in playing condition is credited to Kenneth D. Marshall:

> "*a (violin) was hung on five rubber bands, and a tiny accelerometer was waxed to one spot on the surface where there are as few nodal lines as possible, which in this case was over the bassbar near the G-foot of the bridge. The instrument was then tapped at 190 points by a small, weighted hammer having an accelerometer in its head, the frequency response functions over a range of 0 to 1300 Hz measured with the input force, and the response acceleration was recorded and analyzed to determine the frequency and deflection patterns of the modes.*"
> (Fig. 7)

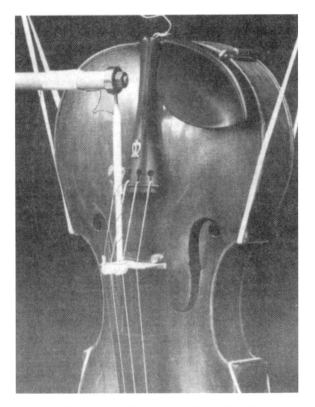

Figure 7: Marshall's experimental configuration for modal analysis of a violin.

Theoretical interpretation of the experimental results were based on the assumption that the violin (exclusive of the strings) could be modeled as a linear second-order system, and its modes of vibrations could be expressed as a summation of the real normal modes of the entire structure. Marshall found *"unsuspected and bewildering variety"* of the vibrational modes of the violin body, a fact vividly illustrated in a modal-analysis animated recording. The recording also revealed the soundpost being *"twisted, stretched and compressed"* by the differential motion of the top and the back at its point of contact.

Marshall's experimental methods of modal-analysis had obvious limitations, as did even those analyses with increasing refinements. The complex design of the instrument, as well as problems with analyzing sound in repeatable experimental conditions, made *it almost impossible to answer the basic question of what happens to violin tones when one changes the thickness, the arching, or other features of the instrument.*

However, the availability of computers and appropriate software brought about a method of analysis based on a completely computer generated model of a violin and ushered violin research into a new era. Known as *finite-element analysis,* violin components were assigned a computer generated discrete mesh structure. Figures 8a and 8b show, for instance, the modeling of the violin top and back plates. With the dimensions and the shape of the plates, the thickness, and elastic properties of each element as input parameters, a computer program obtains the natural frequencies of each plate. G.W. Roberts, using the program called NASTRAN, accomplished the first full finite-element analysis of a violin from free plates to finished instrument (1986).

The input parameters could be varied arbitrarily to discover how much such variations affect the mode shapes and their frequencies. Thus, Roberts studied the effects of changing parameters associated with the plates, such as plate thickness, arching, bassbar dimensions, and edge boundary conditions. Then, in the next stage the two plates were assembled into a complete violin body with ribs, blocks, and sound post. George A. Knott carried a finite-element analysis of the whole violin structure modeled in *vacuo* with free boundary conditions

Cremona Violins

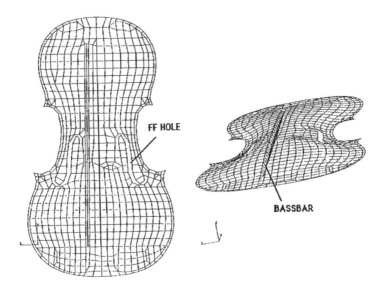

Figure 8a: The mesh structure of the top plate in the finite-element analysis modeling of a computer generated violin using MSC/NASTRAN and PATRAN.

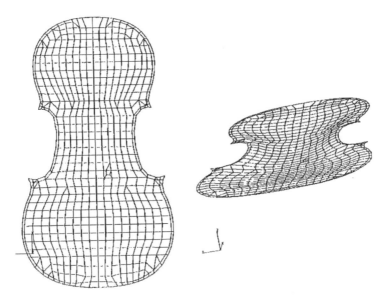

Figure 8b: The mesh structure of the back plate in the finite-element analysis modeling of a computer generated violin using MSC/NASTRAN and PATRAN.

(1987). He developed a system for the violin consisting of 2,220 elements with 10,000 degrees of freedom based on conventional violin measurements and parameters of violin wood. The theoretical shapes and modes of vibrations of free plates and the completed violin could be compared with the experimental results of Marshall and others.

While finite-element analysis aided by powerful computers provided a new and valuable tool to construct a violin and calculate its modes of vibration at various stages of its development, it still proved inadequate in comparison to actual experiments. Such computer generated models, using the dimensions and the material properties in common use by violin makers, provided good "approximations" to real violins, but they could not solve the problem of finding suitable parameters to take into account the effects of internal damping, initial stress caused by string load, and coupling to string resonances, to radiation damping, and to internal air cavity modes.

In recent years, Martin Schleske has developed the application of the computer based modal analysis in a more realistic fashion to obtain what he calls "tonal copies" of any instrument. According to Schleske, when making a new instrument, the guide should no longer be the geometric constructive shape of the reference instrument, but its 'acoustical fingerprint' — the sum of its individual eigenmodes of vibration. Violin should be seen as *"resonance sculpture"* as opposed to a "wooden Sculpture."

In his laboratory, a small pulse generator excites the instrument under study at 600 measurement points. Then, the vibration response is recorded using a miniature sensor. A *"Fourier analyzer"* processes the signals of both the pulse generator and the recording sensor and provides what is called a "transfer function" — the ratio of the vibration response to the excitation force as a function of frequency (Fig. 9). The 600 measured transfer functions are combined using the modal analysis software (called Star Structure) to finally obtain the resonance profile, the *"acoustic fingerprint"* of the instrument to be compared with the recorded tonal quality of the instrument.

Schleske has used this advanced technology to build his own instruments, which are tonal copies of a chosen specific reference instrument. He can also

Cremona Violins

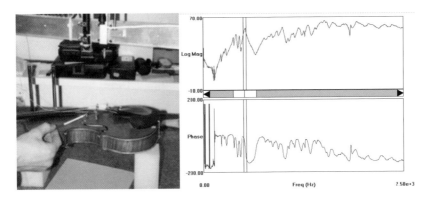

Figure 9: Schleske's laboratory set up; transfer functions.

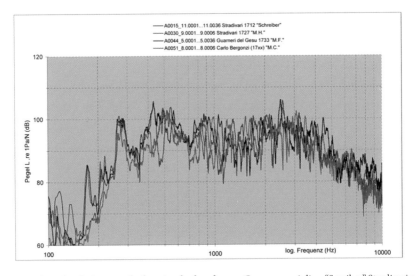

Figure 10: Sound radiation transfer function for four famous Cremonese violins. "Screiber" Stradivarius (red); Stradivarius 1727 (blue); Carlo Bergonzi (violet); Guarneri del Gesu (black).

adjust the modal response at various stages of instrument assembly. Using this technology, Schleske has studied the resonance profiles and tonal copies of two Stradivari instruments (the "Schreiber" anno 1712 and the later Stradivari 1727) and Guarneri del Jesu instruments (Fig. 10). The resonance profiles and the objectively measurable acoustic properties exhibit noticeable differences. However, the exact correlation of such profiles with subjectively perceived

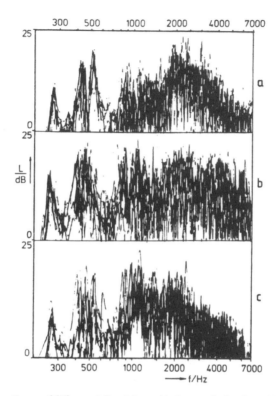

Figure 11: Response Curves of different violins; (a) ten old, distinguished violins; (b) ten violins made by modern craftsmen; (c) ten mass-produced violins.

sound remains a matter of individual judgment. As a further complication, tests in earlier experiments (H. Dünwald, for instance) showed remarkable similarities in such resonance profiles between old Cremona instruments and mass produced ones (Fig. 11).

In Conclusion

With each technological advance, more and more has been discovered about how the violin works, how each component works, and how we can develop theoretical models. However, the essential key, if one exists, to understanding a violin's quality through quantitative characterizations is presently not on the horizon. The perceived sound of a violin depends not only on the physical attributes of the violin but also on the complex interplay between the musician and her/his instrument, the listener, the acoustics of the hall, and so forth. Nonetheless, in spite of these considerations and the controversies and ambiguities in judging the tonal qualities of old Cremona instruments, there is no uncertainty that an experienced violinist can immediately judge whether a particular violin is a student, a professional, or a fine solo instrument. The general consensus in this matter seems to suggest that there is some *objective* quality that can be expressed in terms of a finite set of physical variables. *"But the tantalizing fact,"* says Gabriel Weinreich *"is that no such specification which successfully defines even coarse divisions in instrument quality is known."* Weinreich refers to this frustrating fact as "The New Secret of Stradivarius."

It is the quest for this new secret of Stradivarius that William F. (Jack) Fry has pursued during the past five decades. While strongly motivated toward an understanding how a violin works in the usual mechanical sense, Fry's research

William F. (Jack) Fry

is also directed toward the more difficult, subtle goal of reproducing the tonal qualities of a Stradivari or Guarneri instrument. Committed to a scientific approach and an analysis of the interconnectedness of the various bits and parts of the violin in its physical principles, he looks at the violin as a whole with a "holistic" approach to its acoustics. Convinced that the Cremona masters must have known consciously or unconsciously what they were doing without the technical aid of electronic and other devices, he follows their experimental techniques to achieve predictable changes in an instrument's tonal qualities. With new insights, he has come closer than anyone before him to reproducing the sound of the great Italian violins.

William F. Fry and his Quest for the Secrets of Cremona Violins

The efforts of scientists and those of artists are going to unite to bring to perfection an art, which for so long has been limited to blind routine.

Felix Savart (1791–1824)

William F. "Jack" Fry, professor emeritus at the University of Wisconsin-Madison is well known for his pioneering research in high-energy physics and his work in astrophysics, but during the past five decades he has also pursued research on violins. Fry has been immensely successful in understanding the delicate interconnectedness of the different parts of the violin. His holistic approach to its acoustics, although rooted in sound physics principles, contrasts markedly with the conventional, "reductionist" approach that focuses on separate components, or "mechanical subsystems," of the violin — the bridge, the soundpost, frequency modes, top and back plate resonances, action of the bow, radiation patterns, wood, varnish, and strings. While such research has provided valuable knowledge regarding how a violin works, it has failed to give any clues about what makes the sound of a particular violin distinct from others. Fry's research, on the other hand, is focused on "reproducing" the sound of the great Italian violins. With new insights, he has come closer than anyone before him to achieving this goal.

Jack Fry was born in Carlisle, a small town twelve miles southeast of Des Moines, Iowa, and raised on an isolated farm. Jack's grandfather homesteaded a piece of land of about seventy acres, half of which was woods. Jack's father, with thrift and care, had managed to get the most from the piece of farmland and made enough money to provide a reasonably comfortable living for his family and to send his sons to college. His own formal education stopped at the fourth grade, when he had to quit school to take care of the farm, but he learned enough on his own to become the president of a local bank in his later life.

Jack's mother, however, was a graduate of the City College of Des Moines and a school teacher before she married Jack's father. After her marriage, she devoted herself to taking care of the house, bringing up the children, raising chickens for additional income and helping with the farm.

As a young boy, Jack had two strong interests: music and building radios. *"The isolation of a farm heightened my interest in music,"* says Fry. *"Listening to the radio not only connected me to the outside world, but it also brought music into my life. My ears developed in two ways both at the same time, forming the roots of my scientific and aesthetic interests."* At the age of seven, Jack decided that he wanted to learn to play the violin. His father bought him a fiddle, and he took lessons once a week from a local farmer, an enthusiastic amateur fiddler who played at barn dances and holiday picnics. He taught Jack what he could for two years. When he realized that the young boy had talent, he advised him to receive formal training and directed him to a lady in the neighboring town of Indianola some twelve miles away from Carlisle. Soon Jack was able to play simple melodies quite well and began to play along with his aunt on piano at community events.

In high school, Fry continued to pursue his dual interests in music and electronics, taking violin lessons from whoever he could find and toying with electronics and chemistry in his basement laboratory. The high school had a remarkable music teacher, a Dutchman named Versteeg, whose enthusiasm for music far exceeded his talents. *"He played one instrument, the trumpet, badly,"* says Fry. *"He knew the notes on the piano and could tune a piano but was not in any way an accomplished pianist."* However, because of him, the school of only ninety students had a band and an orchestra of forty-five. He was an inspiring head whose encouragement and discipline led his group to perform, and perform well, winning contests in county and state competitions.

Fry was his most versatile and talented student. Versteeg gave him a free hand in the orchestra to teach and direct other students. At the same time, Fry was also interested in physics. But he learned physics on his own, since his physics teacher was an athletic coach who knew very little physics. Fry not only taught himself, but he came to the rescue of his teacher and was often asked to teach other students. All this gave Fry enough confidence to take on a dual major in music and electronics when he joined the Iowa State University in Ames, Iowa in 1939.

In his freshman year, Fry had the opportunity to take violin lessons from a true professional, Ilse Niemack, a contemporary of Jascha Heifetz. She had been trained in many European music schools and was an accomplished violinist. Her standard of playing was far above anything he had encountered before. Under her strict but wise and inspiring tutelage, he learned advanced techniques and developed a deeper appreciation for classical music. At the same time, observing and listening to Niemack's playing, Fry became convinced that he lacked the combination of will and talent required to become a professional musician. He decided to concentrate on his scientific interests and to pursue a career in engineering.

After successfully completing his undergraduate degree in engineering, Fry spent four years (1943–47) at the Naval Research Laboratory in Washington, D.C., involved in war-related work. He built radio jamming devices and a huge transmitter to misguide the radio-guided German missiles. While eminently successful in his engineering work, Fry soon realized that his real interests lay elsewhere, not in building things, but more in understanding the principles behind them. Thus he decided to make another turn in his career; he took night courses in physics at George Washington University, where as luck would have it, he was able to study with George Gamow, not only one of the foremost physicist of twentieth century science but also one of its most charismatic figures. Under his spell, Fry discovered physics as his true vocation. He studied atomic, nuclear and theoretical physics and was admitted to graduate school at Iowa State University. His graduate research on high energy cosmic rays won him a postdoctoral fellowship from the Atomic Energy Commission. After a year at the University of Chicago on this fellowship, he joined the physics faculty of the University of Wisconsin-Madison in 1951, where he established himself as one of the pioneering researchers in high-energy physics.

Fry had long ago given up the violin except to occasionally play in friendly amateur quartets or to accompany a pianist. Then, one evening in Berkeley in 1961, Wilson Powell, a colleague in research, suggested they play some music. Since Fry had no violin with him, Powell borrowed two from the music school, a Stradivari and a Gagliano. For Fry, it was a momentous revelation. *"For the first time in my life,"* says Fry, *"I realized how a good violin can change one's ability to play. That evening was like an awakening. Here were two instruments which were so different, and so much better than anything I had ever played on. No one ever had pointed out to me that some of the problems of playing the violin are related to the problems of the violin itself.*" He became fascinated by the old instruments. How could they be so different from the violins he had played? Was there a scientific explanation for the phenomenon?

On his return to Madison, in search of a better violin for himself, Fry went to the violin shop of Larry Lamay, who repaired as well as made violins. He was a skilled craftsman and had made some good instruments but was keenly aware that they did not match the sound of some of the famous Italian instruments he had come across. Like Fry, he was also interested in solving the mysteries of the great violins of the past, and he strongly believed that the secret lay in the *varnish*. That was not surprising for Fry. For centuries people had been indoctrinated with the idea that some simple thing, one simple characteristic, was the clue to a great violin, and *varnish* has been the main focus of this idea. "*It is not unreasonable that people think* ***varnish*** *is the secret,*" says Fry, "*because if you play an unvarnished violin and then play it again after varnishing it, you notice dramatic changes. It is one of the obvious things you can use to demonstrate that it does change the acoustics of the violin.*"

Lamay allowed Fry to varnish several of his instruments, and, indeed, Fry discovered that hardening the varnish and increasing the drying temperature changed the sound. It made it harsher but louder. He thought that he had made a breakthrough and spent the next one and half years experimenting with varnish. He discovered several things about varnish, several ways to change the varnish and apply it to produce noticeable acoustical effects. A thin varnish will soak into the wood more. Alternately, one can increase the amount of varnish that gets absorbed into the wood by putting on multiple coatings. One can treat the wood, for instance, by soaking it in water and allowing it to dry. That tends to open the pores and allow more varnish to be absorbed. Varnishes polymerize, imparting strength to the wood fibers. This can be controlled by the temperature at which the varnish is allowed to dry. "*I did a lot of crazy experiments,*" says Fry. "*I tried impregnating woods with various substances and varnishes and heating the wood to high temperatures to dry it out or shrink it.*" But soon he found that such experiments led nowhere. The results were totally unpredictable. The same process on two different violins led to entirely different results, good at times and sometimes so bad that the violin was essentially destroyed. Lamay had to make a living, and it was not easy for him to see a fair number of his instruments ruined by Fry's experiments. Besides, like most violin makers, he also wanted to keep their explorations secret for commercial reasons. However, that did not suit Fry, whose primary aim was a scientific quest.

At this point, Wilson Powell, who had been following Fry's efforts closely, came to his rescue by offering him unconditional financial support. *"Don't worry about making a great violin,"* Fry recalls him saying. *"You do your research. Do what you want to do. Buy any number of white violins and send me the bills. I will take care of them."* This was the beginning of Fry's serious research on violins. He left Lamay's workshop and started working independently, working with unvarnished (white) violins, which would cost between $50 and $200 each. If they turned out well, he would sell them for a profit and put the money into a corporation he formed with Powell to cover expenses and buy more instruments for experiments. Fry spent two more years experimenting with varnish before he convinced himself that varnish alone could not be the mystery behind the great Italian violins. Furthermore, he became frustrated with the fact that it took six or more months for the varnish to harden and attain stable acoustical properties enough to be tested. He needed another approach.

Through his experimentation, he realized that the main effect of varnish was to change the elastic properties of the wood. Wood is naturally strong along the grain and weak in directions transverse to the grain. Since varnish penetrates into the wood, its polymer properties strengthen the fibers in the transverse directions. The net result is a change in the stiffness. Since the degree to which varnish can penetrate depends on the inherent properties of the wood, and no two pieces of wood are identical, there is no way to control and duplicate the stiffness variations from one instrument to another. On the other hand, since stiffness is a very sensitive function of thickness, it is infinitely easier to compensate for change in stiffness due to varnishing by varying the thickness of the wood.

With this insight, Fry began to experiment with varnished nineteenth century "old junk violins," which he found in antique shops and violin stores. In such older instruments, he often found the craftsmanship to be better than in new instruments and also the wood more stable against climatic changes. On such instruments, by scraping or placing inlays in appropriate places, he could control the changes in the acoustical quality and work towards producing a desirable timbre. He invented a simple device with which he could reach through the f-holes to any part of the interior surface of the violin plates and scrape and alter the thickness variations. He was certain that the great violin makers must have employed such a device in order to fine tune their instruments in the final stages of construction.

It was common knowledge among the violin makers that there were subtle thickness variations or graduations in the top and back plates. It was noted that the back plate was thicker in the central region and thinned out at the extreme ends. The top plate was thinner and more uniform. In addition, there were regional variations in the thickness graduations of the plates. But little importance was attached to these features. For instance, in *Antonio Stradivari: HIS LIFE & WORK*, we note measurements on several instruments, including those of Andrea and Nicolo Amati and Stradivari, showing a typical thickness of 10–14/64 inches at the center graduating down to 5–7/64 at the flanks. The authors conclude, *"Neither do we find that Stradivari — or, in fact, any of the great makers — sought to obtain absolute precision in the working of the thicknesses; the whole is carefully wrought out, yet without any attempt at mathematical exactness. They were apparently of the opinion that greater precision was unnecessary."*

Likewise, Felix Savart (1791–1841), a brilliant physicist who was the first to apply scientific methods to the study of the acoustics of stringed instruments, had made observations and experiments on the best Italian violins, particularly those of Stradivari. He had noted the graduations of the top and back plates and found that the thicknesses almost always varied even in violins made by one maker using wood of the same quality. However, the instruments retained the distinctive character of their sound, leading him to conclude that the thickness variations were not important. From his measurements, he deduced some average variations that were totally symmetrical around a vertical axis through the center of the plates that were to become standard in Vuillaume's instruments (Figs. 1–2). Many subsequent violin makers followed this example. It became a common belief that plates should be worked out to the millimeter thicknesses and blend areas of different thicknesses so that there are no hills or valleys but only a uniform increase or decrease.

For Fry, however, the thickness distribution pattern was an extremely important consideration in the mechanics of sound production in a violin. As discussed in Chapter III, he conceived three important modes of vibration associated with different frequency ranges. In the low frequency mode, the **Breathing Mode** dominates, and in this mode, the top and back plates move out of phase, pushing the air in and out through the f-holes. If this is to work

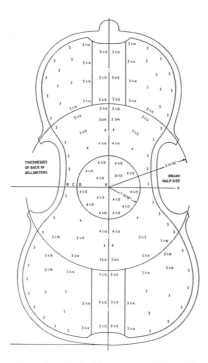

Figure 1: Graduations of the back plate that had become standardized showing perfect symmetry about a vertical axis through the center.

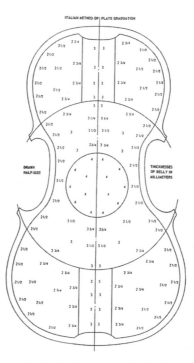

Figure 2: Graduations of the top plate that had become standardized showing perfect symmetry about a vertical axis through the center.

efficiently, the top and back plates have to move in a piston-like fashion *without the plates undergoing significant rocking or rotational motion*. Fry determined that symmetrical variations in thicknesses of the type shown in Figs. 1–2 will not meet those conditions since the driving force, *the soundpost*, is not at the center of the plate. In order to compensate for this off-center location of the driving force, thickness variations around the center have to be *asymmetrical*. Using simple principles of physics, Fry deduced the shifts of the uniform thickness contours away from the symmetrical configurations (Fig. 3).

To test these ideas, however, he needed to measure the thicknesses of the plates. It was unthinkable to take apart a Stradivari or a Guarneri just to make such measurements. Once again, Wilson Powell came to his rescue by inventing a simple device. It consisted of introducing a tiny magnet through the f-holes at a desired location inside the back plate and another magnet

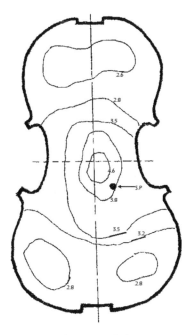

Figure 3: Fry's asymmetric graduations in the back plate expected from theory.

placed on a "postage stamp" weighing balance just underneath the one inside. The magnets would attract and would be held in place. When the violin was slowly lifted, the magnets would disengage and the spring in the balance would be extended. By suitably calibrating it with wooden strips of uniform thickness, Powell was able to measure the thicknesses at various points of the plate (Figs. 4–5).

With this device in hand, Fry contacted the late Leonard Sorkin, first violinist in the Fine Arts Quartet at Milwaukee. He played on a Guarneri Del Gesu and was quite receptive to Fry's scientific ideas. Powell and Fry measured and obtained the contour maps of the back plate and to their delight found the expected asymmetries (Fig. 6). Then, they obtained contour maps of several good instruments. This work became so important to Powell that he continued to measure graduations in the top and back plates of violins. He found the expected asymmetries essentially in all the Italian instruments (Fig. 7); in other instruments, they were generally absent or if present, were neither of the right character nor of the right magnitude. At the same time, the observed asymmetries in the Italian instruments, although they were of an expected order of magnitude, were not exactly such as to completely prevent the rocking or rotational motion of the plate. They left open the possibility of some combinations of low frequency rocking resonant modes that gave distinct voice to the instrument.

Fry spent the next couple of years experimenting with asymmetries in thickness variations in the plates. While such experiments with "junk violins" led to tremendous improvement in sound quality, lending them a deeper and

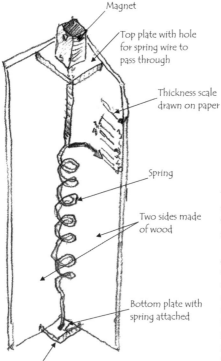

Magnet

Top plate with hole for spring wire to pass through

Thickness scale drawn on paper

Spring

Two sides made of wood

Bottom plate with spring attached

The spring length was adjusted so that when the magnet sat on the top plate there is little or no tension in the spring.

An equal Magnet was dropped into the violin through the f hole. A strong magnet from outside the violin wall used to move the inner Magnet to where the measurement was to be made. This also oriented the small internal Magnet so that the S pole was against the wood. Someone held the violin still while Powell brought his device to below this position raising the measuring deviceupword to the violin surface. The Magnets would "click" together. Powell would then slowly lower his device which would strtch the spring until the force of the spring would separate the Magnets. This breaking point would then be read on the previously calibrated scale.

Figure 4: Wilson Powell's device to measure the graduations.

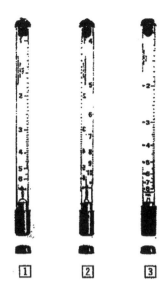

HACKLINGER Gauge

This unique tool allows exact thickness measurements of stringed instruments to be made **without disassembly**. The small counter-magnet is passed through the instrument's f-hole and follows the gauge around every edge and curve, allowing the exact thickness to be measured at any place on the body. Accurate to within 0.1 mm, calibrated for making measurements while holding the gauge in a vertical position (horizontal use will result in a measurement error of +3%). Quick and easy to use, with no risk of damaging even the most sensative varnishes. Ideal for the study of old instruments, locating old crack repairs, making sound adjustments, etc. Also suitable for measuring other, non-ferromagnetic objects (e.g. boat hulls, plastic containers, ceramic vessels, etc.). Comes in a leather sleave with instructions for use.

Sole German distributor.
(Patent No. 3611798)

Figure 5: Commercial device that uses Wilson Powell's magnet idea. No credit is given to Powell.

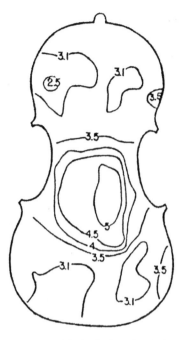

Figure 6: Wilson Powell's measurements of the back plate of Sorkin's Guarneri violin, confirming the asymmetries.

Figure 7: Wilson Powell's thickness measurements of plates of a number of famed Cremona violins.

augmented voice in the low frequency range, none of them turned out to have the quality of sound of a Stradivari or a Guarneri or that of even less renowned Italian instruments. To Fry's trained ear, it became clear that asymmetries in the graduations of the plates were just another set of parameters but not the whole story. Still in search of the single most important acoustical response that made a violin great, he spent the next 2–3 years studying the frequency response of a violin.

To measure the frequency response, Fry attached a transducer to the bridge and drove it by applying varying frequency inputs and using a microphone to pick up the radiated sound. As faithfully as possible, the process imitated the actual vibrations of the strings of a violin that make the bridge move. He measured the frequency response of dozens of instruments, some of good quality and some poor, and found very little to distinguish one from another based on their frequency response. He found the responses extremely complicated involving literally hundreds of resonances with little correlation between the frequency response and his perception of the quality of sound. Although the technique he used to measure frequency response was rather crude, measurements with more refined apparatus using oscilloscopes and other techniques have led to the same conclusion (see Fig. 11 in Chapter 4). Indeed, such an analysis reveals an extremely complex pattern of sound containing an enormous number of frequencies with little or no correlation to quality of sound. *"I kept on telling myself,"* says Fry. *"Stradivari never had such electronic devices. Still he made such great instruments. Why not use his techniques?"*

Fry became convinced that frequency analysis was not the right approach. Fry says, *"It took me a long time to get over the idea that there were one or two simple secrets. It is much more complicated than just measuring the frequency response or some other isolated physical characteristic. The dynamic properties of the instrument are as important as the sound character of the instrument, and the two are strongly coupled. We are dealing with a frequency range from 220 Hz to 12,000 Hz or higher and rapid changes from one frequency to another as the violin is played. The motion of the box is obviously very complicated. I began to think of models of how to simplify the understanding of the way the box moves and what makes it move the way it does."*

Before we go into details of the models Fry developed during the course of his research, it is important to note two conceptual advances Fry made in his scientific approach to the acoustics of the violin. The first was the recognition that the individual sharp resonances did not determine the tonal quality of the sound of a violin. Rather, the collective perception of a group of frequencies by the ear signals the specific timbre of a violin. This is in strict analogy with human speech. The ear, a complex analyzer of sound, distinguishes one voice from another based not on one or two frequency components but on groups of frequencies called *"formants"*. In the case of the violin, Fry recognized and named three formants, each linked with a predominant mode of vibration of the violin. As described in Chapter 3, these are; 1) The low frequency range (200–1000 Hz), *the breathing mode*, 2) The mid-frequency range (2000–5000 Hz), *the rocking mode, and* 3) The high frequency range (5000 Hz and higher), *the tweeter mode.*

The other important advance Fry made was to isolate certain *acoustical absolutes* that are essential features of a good violin and finding how they are related to the mechanics of sound production. Clearly the first important characteristic of a good violin or any violin for that matter is its *carrying power*, the ability to be heard over other instruments even when they put out more energy. This attribute depends upon the ability of the violin to radiate high frequencies efficiently (around and above 6000 Hz) with as much power as possible. This is because other instruments by and large do not emit the harmonic content in the high frequency range with enough energy. A horn, for instance, can put out an enormous amount total energy compared to a violin, but its output is minimal in the 10 kHz range.

Besides the carrying power, efficient radiation in the high frequency range has other advantages. The high frequency component has, what is called, a *silky* or a *refined* quality as opposed to a *shrill* sound. The latter is associated with the mid-frequency range around 2.5 kHz to which the human ear is most sensitive and generally causes an unpleasant sensation. Moreover, it is a remarkable feat of the human brain that, even when the fundamental is not there, it has the ability to reconstruct and perceive it. So the listener is sub-consciously aware of the note being played by the violin! This perception of the fundamental adds depth to the silky quality of the high frequency radiation.

A good violin, for any single note played, has a *divided sound*, meaning the low frequency range containing the fundamental is well separated from the high-frequency range with the shrill mid-frequency range suppressed. The low-frequency component gives the listener a feeling of pitch and fullness of sound; the high-frequency component gives a feeling of elegance, fineness and a very pleasing sound. A violin that has no such separation between low and high frequency components and emits a sort of continuum note is said to be *loud and shrill*. It lacks the elegance and fineness of the divided sound.

Even and *uneven* from note to note and *Wide Dynamic Range* characterize two other absolutes of a good violin. If the intensity changes enormously as a function of frequency, the sound is *uneven* and certain notes may stand out. On an instrument where certain notes stand out, the player has to produce *evenness* manually by playing some notes harder or softer than others. A violin with *wide dynamic range* can be played very softly without losing its character and at the same time allows the player to put more pressure on the bow to increase the volume without affecting the fundamental properties of the sound. A good violin should also allow *flexibility* so that a violinist can project feelings and emotions by varying texture and by changing from a pure sound to sometimes even a rough sound. If an instrument does not allow this flexibility, it sounds "frozen," producing always the same type of sound no matter how it is played — softly or loudly.

Violin *response* is a complex subject, but as the word suggests, it concerns how quickly a note can be excited. A good violin responds almost instantaneously; the note begins as soon as the finger touches the string, as though, a violinist might say, *"It anticipates the note I am going to play; the note seems to start before I play it. I do not have to 'attack' the note."* Another aspect of *response* concerns how a violin sounds to the ear of the player. The "local" sound surrounding the box and hence near the ear of the player is dominated by low frequencies, whereas the sound radiated out has more of the high frequency components. If the local sound is too loud, the player will tend to play gently to produce the high frequencies that carry out to the audience. If it is softer, he will be more forceful and the increased intensity of the high frequency components results in greater carrying power.

Finally, *ringing*, or the continuation of the sound when the bowing slows or stops, is one of the most important absolutes from the point of view of both the player and the listener. Human perception of the sound frequency, *the pitch*, depends sensitively on the frequency, the intensity of the overtone structure, and the duration of the sound. An open string has a natural ringing lasting approximately 2/10 of a second during which one has a strong perception of the fundamental pitch. But as the intensity of the ringing decreases with time, one perceives a change in the *timbre* of the sound although there has been no change in the frequency of the emitted sound. The high frequency components lose out in intensity faster than the middle frequencies. In general, the pitch perception of lower frequencies increases with decreasing intensity. The opposite is true for high frequencies. The perception of the shrill region, however, is independent of intensity. So ringing has a strong effect on the timbre of the notes one hears.

The second type of ringing called *resonance ringing* is perhaps one of the most vital components of violin acoustics. It has to do with the fact that every note on a violin string has the potential of exciting a high overtone on another open string. The best example is the excitation of a high g-note on the D-string when a g-note is played on an open G-string. As a result, even when the playing on the G-string stops, one continues to hear the g-note because of the D-string. Clearly such couplings between strings occur through the bridge and give distinct timbre to each individual note. If the coupling is not right or the violin is not tuned right, there is no such resonance ringing and the timbre of each note undergoes a noticeable change. A musician feels this change and uses it to monitor the frequency. Another important aspect of this type of ringing is its effect on the perception of *vibrato*, which is the small oscillation in frequency around a specific note one is playing. If a violin is tuned right and if strings are properly coupled, vibrato on one string excites overtones on another string and leads to distinct change in the timbre. The ear uses this change in timbre to detect the small change in pitch.

As opposed to this highly desirable *resonance ringing*, there is the possibility of *false ringing* that occurs when the sound of a particular note on a particular string persists even after the bow has passed on to play another note. This is not desirable since it affects the timbre of other notes, whereas resonance ringing or true ringing connects notes on different strings and provides continuity during transition from one note to another.

These acoustical absolutes have little to do with the physical appearance of the instrument, although most of the Cremona violins are extremely beautiful. Stradivari, for instance, crafted some extremely beautiful instruments from the point of view of their outward appearance — the way the varnish is shaded, the way the purfling is placed with such perfect skill and regularity and done with right contrast of woods. These are the most talked about features in violin auctions and exhibitions. However, Fry is strongly convinced that there is very little correlation between the outward appearance and outstanding acoustical properties. He cites the example of Carlo Giuseppe Testore (c.1660–1720), whose cheap and unfinished instruments, still bearing scraper marks, are as beautiful acoustically as are his beautifully finished instruments. There are other similar examples of instruments made by Stradivari, Guarneri and others. Leonard Sorkin, for instance, owned a Stradivari and played on it during his early years. Its back had no flames and was *"as plain as the board on a kitchen sink,"* according to Fry. *"Its top plate had the indications of a knothole so that one would think the choice of wood was not particularly suitable for a violin. Yet, the instrument had a beautiful voice."*

Fry believes that a large number of Cremona instruments are in a class by themselves. They are unique with respect to the acoustical absolutes when compared with the instruments made elsewhere. This can be easily tested, Fry says, by simply drawing the bow at constant speed across the string with no dynamical modification. The sound from old Italian instruments is recognizably different from that of other instruments. At the same time, within the class, violins of great masters have their own distinct voice. A violinist might say, a Stradivari sound is more **refined** and **delicate** than that of Guarneri. The latter has a deep enchanting **robust colored** voice, but it is **frozen**. For one player, a Stradivari may be too delicate to be played forcefully and hence lacks in dynamical range. For another a Guarneri sound may be too frozen, making him unable to change the **color** of the sound to express his varying emotions.

Given the complexity of the instrument, the interaction between the player, and finally the perception of the listener, it appears forbiddingly difficult, if not impossible, to provide recipes scientific or otherwise to construct instruments with any degree of predictability. Yet Fry has been able to come close to this goal by departing from certain standard approaches and finding new ways to handle the complexity.

The most significant idea behind Fry's research is the realization that the violin is a driven system, a combination of the vibrating string as the generator providing a wide frequency range and its harmonics, and the sounding box as resonator, responding selectively according to its own pattern of resonances. It is the box that amplifies and radiates. This radiated sound characterizes the quality of sound and the personality of the instrument. For instance, the sound of a great Stradivari instrument is extremely silky, that is, the high frequencies in the range above 5000 Hz and above are exaggerated, which gives it a brilliance and refinement. Whereas a Guarneri is, what a violinist would say, more colored in its sound, meaning it has more of the fundamental compared to the higher frequencies.

The resonances of the top and back plates and the natural resonances of the enclosed air are not free. They are excited by the harmonic forces generated by the vibrating strings and communicated through the bridge. Hence, their response depends upon how they couple to these forces. According to Fry, the human voice provides an apt analogy. The vocal chords drive the resonating system of the lungs, the nasal and the oral cavities. Each of the latter has its resonant modes. However, these individual resonances don't distinguish one person's voice from another. Instead, a distinct voice depends on how the vocal chords couple to and drive the resonating cavities.

The realization that the violin box is a driven system led Fry to visualize it in terms of an idealized mechanical system with the top and back plates as a set of springs and masses, the spring constants and mass values to be determined by the local thickness graduations of the plates. While the details of the model are deferred to the next chapter, we note here that the spring constant is proportional to the cube of the thickness, while mass is just proportional to the thickness. By varying the graduations, one can vary both of these parameters. The result is a system of what a physicist would call a system of coupled oscillators driven by external harmonic forces. Carried to its ideal limit, it becomes a system of tremendous complexity. A rigorous mathematical analysis, while not impossible, is of little use in making a violin.

For Fry, the problem was a practical one; it was to find the parameters that couple the three basic modes and find ways to change and control them in order to play the modes independently with desired amplitudes. For instance, if a violin sound is too soprano, and a player wants it to sound deeper, the amplitude of the breathing mode has to be increased. If it sounds shrill, the rocking mode has to be suppressed. To increase the carrying power and the elegance of the sound, the high frequency range has to be enhanced. Since the modes are so strongly coupled, changing one requires changing something else. From numerous experiments, Fry has been able to accomplish this to reach higher and higher levels of predictable sound quality.

In this semi-intuitive approach, backed by sound principles of physics in constructing models and experimenting, Fry had the benefit of interaction with a number of violinists over the years. Among them, according to Fry, Rosemary Harbison, an accomplished and highly acclaimed violinist, has been an "untiring contributor" to the understanding of the complex world of violin sound. There is an enormous variety of sounds of different qualities in Cremona instruments, instruments such as those of Amati, Stradivari and Guarneri. No one instrument sounds exactly like another although made by the same master. Yet, each one of them has the distinct voice of their maker. How then can one attribute a unique quality and preeminence of one characteristic over another pertaining to an instrument? This raises many complex issues. Fry needed the help of a professional violinist to play and react and "tax" the instrument to its limit to get its response. Rosemary's critical analysis and constant guidance has been of utmost help for Fry in his research.

6

William F. Fry and his Quest for the Secrets of Cremona Violins
(*continued*)

Stradivari never had any of the modern electronic devices. Think of the great instruments he produced. Why can't we reproduce them with the techniques he used? He had two major tools: A set of calipers that could measure thickness accurately to tenth of a millimeter and the other, a scraping device to modify thickness graduations once the violin was completed.

William F. Fry

In the absence of any modern technical devices, it is reasonable to assume that Stradivari, his predecessors and followers had to rely on their ears and senses to accomplish their desired goals. Their method could only consist of an empirical selection of what they had learned to be the most advantageous combinations of woods, design, arching and the graduation and flexibility of various parts of the top and back plates. Fry's research is in this old tradition. He has, of course, the advantage of deep insights based on sound physics, whereas most likely, the Cremona violin makers relied on their intuition, craftsmanship, experimentation and knowledge passed on from one generation to the other.

An important aspect of Fry's research is based on the well known fact that the response of the completed instrument has little or no relation to the unassembled pitches of the plates. Whatever the frequencies to which the free plates may be tuned, varnish and gluing together would invariably alter the final response. Hence, for Fry, there had to be a method and a device to control and change the response of the instrument after it was completely assembled. Since his research involved experiments on completed instruments, he needed such devices to measure the thickness graduations and alter them as necessary without opening the violin box. Wilson Powel had provided the thickness measuring device. To make the alterations, Fry invented a simple homemade "scraping" device (Figs. 1a–1b) that could be inserted through the f-holes and reach the interior parts of the plates and change the graduations. To his great delight, later during his visit to the Stradivari museum in Cremona, he found that Stradivari did indeed have such a device.

The basic idea underlying Fry's violin acoustics is that the violin box is a driven system. In such a system, the response of the violin will depend upon how the different modes are coupled and mixed and how smoothly and continuously they go from one mode to another. There are basically two parameters at Fry's disposal, namely, stiffness and mass, both depending upon the local thickness distributions. Stiffness (technically characterized by Young's Modulus) of wood is proportional to the thickness cubed, whereas mass is proportional to thickness. For Fry, most of the critical acoustical problems are associated with the top plate. The back plate plays an important role in the

Figure 1a: Stradivari's set of calipers that could measure thickness accurately to tenth of a millimeter.

Sketches to scale of three scrapers

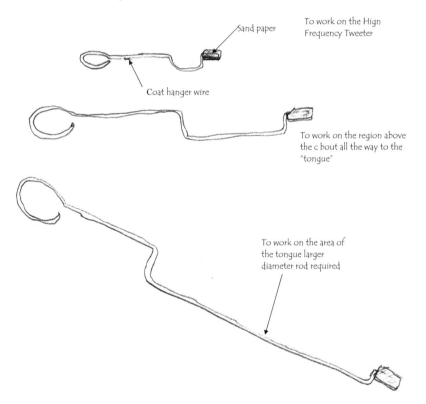

Sand paper

To work on the High
Frequency Tweeter

Coat hanger wire

To work on the region above
the c bout all the way to the
"tongue"

To work on the area of
the tongue larger
diameter rod required

Figure 1b: Fry's scraping devices that could be inserted through the f-holes and reach the interior parts
of the plates and change the thickness graduations.

low frequency range (200–2000 Hz), ***the breathing mode,*** but as one goes up in frequency, its role diminishes. Maple used for the back plate has a higher density than spruce used for the top plate. The higher density and its greater thickness give the back plate enough inertia to remain almost stationary at higher frequencies. The high frequency motion and radiation are virtually entirely controlled by the top. Therefore, the art and science of Fry's research consists in isolating certain critical areas, mainly of the top plate, and finding a system of graduations such that the resulting driven system functions as desired in different frequency ranges. These critical areas are illustrated schematically in Fig. 2.

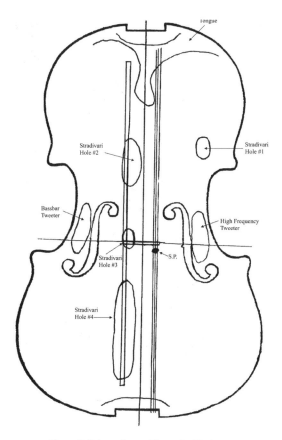

Figure 2: Schematic markings of critical area.

Looking down at the top of the violin (as held in the usual playing mode), we note that the left foot of the bridge rests on the *bassbar*, a beam of constant width, high mass and variable thickness. It is thin at the ends and thick at the center. Its center of mass is displaced towards the scroll from the point where the foot of the bridge meets the plate. On the opposite side, the right foot of the bridge rests over what Fry calls *soundpost fibers (SPF)*, which extend over a width approximately equal to the diameter of the *soundpost*. As with the center of mass of the *bassbar*, the *soundpost* is also displaced from the foot of the bridge towards the chin (Fig. 3). The *bassbar* provides structural reinforcement of the top plate on the side opposite the soundpost. Its important properties are its mass, its length and its shape. It is linked with two types of motion, the displacement and rotation.

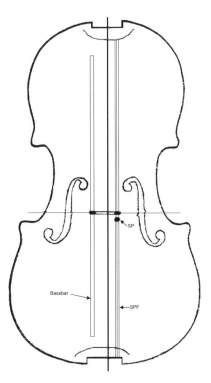

Figure 3: Seen from the top, pictures of bassbar, soundpost fibers, and the positions of the feet of the bridge relative to the center of mass of bassbar and the soundpost.

On the other side, the *soundpost fibers (SPF)* are long and weakly supported, since they are only attached to the two end blocks. They are strong along the grain (along the length) but weak in the transverse direction. Because of this large asymmetry in stiffness, it is reasonable to consider this group of fibers by themselves to understand their special role. "*Soundpost fibers are the nervous and muscular system of the violin*," says Fry. The efficiency in their functions in the three principal modes requires a complex system of support that underlies the areas schematically shown in Fig. 2.

The low frequency range, the *breathing mode*, endows the violin with its deep sound response due to the fundamental frequencies of G, D, and A strings. There are three ways in which the plates move in the low frequency range: (a) A piston-like displacement motion in which the two plates move in opposite phases. (b) Rotation of the plates around the central horizontal axis (X-axis); (c) Rotation around the central perpendicular axis (Y-axis). The predominant motion of the plates in the breathing mode should be a piston-like, letting the box 'breathe in' and 'breathe out' the air through the f-holes.

If we concentrate on the back plate, we find that its motion in this mode is controlled by its stiffness and not so much by its mass. Mounted alone without the soundpost, it has its own natural resonant displacement and rotational modes. However, with the soundpost, it becomes a driven system, and its inherent frequencies become irrelevant. The response then depends upon the thickness graduations in two areas, near the geometrical center of the plate (centroid) and around the soundpost. The maximum thickness, varying in the range 4.8 to 3.5 mm, is to be found near the geometrical center of the plate, not necessarily near the soundpost. The asymmetrical location of the sound post requires, as discussed in the previous Chapter, proper asymmetries to augment the piston-like motion. The overall low frequency personality of the violin depends sensitively on the details of the thickness graduations of the back plate. For instance, Guarneri instruments are perceived to have a greater depth in the low frequency region compared to those of Stradivari. This can be attributed to the fact, as Wilson Powel's measurements show, that the maximum thickness occurs, almost without exception, at the geometrical center in the case of Stradivari instruments; displaced to the left and above the soundpost. However, in the case of Guarneri instruments, it is closer to the sound post. This drives the piston-like motion harder than the rotational modes, resulting in the sound post being driven more aggressively, adding more depth to the low frequencies.

How does the top plate function in the **breathing mode**? In the frequency range under consideration, if we neglect the displacements of the center of mass of the *bassbar* and the position of the soundpost from the foot of the bridge, we expect the *bassbar* and the *soundpost fibers (SPF)* to move in opposite ways (phase) as the bridge rocks from left to right. When the left foot of the bridge pushes on the top plate, the *bassbar* goes down and the *SPF* go up. The reverse happens when the right foot of the bridge comes down on the plate; the *bassbar* goes up and the *SPF* go down. However, a more careful analysis shows that, some of the time, the *bassbar* and the *SPF* are in phase in the lower bout of the violin while they are out of phase in the upper bout. This is because of the displacement of the soundpost from the foot of the bridge. As the right foot of the bridge pushes on the plate, the *SPF* rotate around the horizontal axis so that the *SPF* go down in the upper bout while the *bassbar* goes up. This phase difference makes the right-hand side of the upper bout cancel the air displacement on its right-hand side. The net result is that the piston-like motion of the entire top plate is impeded. The role of the *tongue* in Fig. 2 is to prevent this from happening. Its primary purpose is to reduce the rotation of the *SPF* to the maximum extent and enhance the piston-like motion, thereby augmenting the **breathing mode**.

The shape and the thickness graduations of the *tongue* are shown in some detail in Fig. 4. We note that the region around the neck block has maximum thickness extending to the right over to the *soundpost fibers (SPF)*. As it goes down towards the f-holes, it slants and passes very close to the *bassbar* before turning and going very close to the *SPF*. The bottom of the *tongue* is located just below the maximum bulge of the upper bout, which occurs half way between the C-belt and the neck block. The design of the *tongue*, with thickness graduations shown in Fig. 4, is such that it effectively acts like a lever

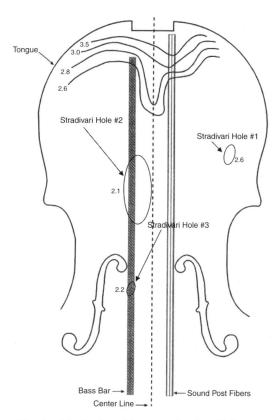

Figure 4: Details of the sketches of the tongue, Stradivari Holes #1, #2 and #3.

between the **bassbar** and the **SPF**. The **tongue** where it is close to the **bassbar** follows the latter's motion, magnifies it and communicates it to the **SPF** at the other end so that the **bassbar** and the **SPF** move in phase. This is further facilitated by thinning the **SPF** just below the **tongue** and making it the point where the **SPF** flex. The net effect is that the **bassbar** and the **SPF** move in phase in both the upper and lower bouts, nullifying the cancellation of the air displacement in the upper bout.

An equally important effect of the **tongue** is to shorten the length of the **soundpost fibers (SPF)** by the length of the **tongue**. Their effective length is essentially reduced by the length of the upper bout. Since the **SPF** are very flexible compared to the **bassbar**, this shortening stiffens them around the vicinity of the *soundpost* and this in turn reduces the tendency of rotation around the *soundpost*. Consequently the right foot of the bridge drives the *soundpost* harder and adds depth to the sound in the low frequency range. Appropriately graduated, the *tongue* is thus responsible for an increase in the loudness of sound in the *breathing mode*. As explained earlier, this is accomplished by preventing the cancellation of the air displacement in the upper bout by making the *bassbar* and the *SPF* move in phase. Secondly, by shortening the length of the *SPF*, it diminishes the rotation of the *SPF* and drives the *soundpost* harder causing the increase in the depth of the sound. This diminishing of the rotation of the *SPF* serves another extremely important purpose, and that is the suppression of what is called the Wolf-note in the literature. This occurs in the frequency range of the *breathing mode*.

The rotation of the *SPF* communicated to the *bassbar* causes the top plate to rock around the horizontal axis, which in turn makes the air in the box slush back and forth. The box acts as a closed pipe resonance, the frequency of which can be calculated from the known standard length of the box. This natural resonance frequency of the box is found to be at the pitch of the B-flat on the A-string (an octave above the B-flat on the open G-string). In many violins, this note is so dominant that this particular note on the G-string cannot be played; the string refuses to excite the note. It flutters and skips from below to above it in frequency, since it takes too much energy to excite the note at the resonance.

As frequency increases to the mid range of 1500–4000 Hz, one encounters the *shrill region* where the human ear is most sensitive. It gives rise to an unpleasant sensation that has to be suppressed. From dimensional arguments, it is easy to estimate the area of the radiating surface in this frequency range. It turns out to be roughly the area of the top plate. It is the rocking motion of the top plate that is responsible for the radiation in this frequency range. This rocking motion is due to the rotation of the bassbar. At very low frequencies, the *bassbar* undergoes mostly vertical displacement because of its mass, but its rotation begins to be significant as we approach the upper end of the low frequency region.

To suppress the shrill region, it is therefore necessary to minimize the low frequency rotation of the *bassbar*. Its rotational motion is caused primarily because of the displacement of its center of mass away from the foot of the bridge (see Fig. 3) and secondly because of the coupling between the *bassbar* and the rotational motion of the *soundpost fibers (SPF)*. The latter rotate because of the displacement between the *soundpost* and the foot of the bridge on the right side, the side opposite to the *bassbar*. The *SPF* rotate with the *soundpost* acting as the fulcrum, and they communicate their motion to the *bassbar* and enhance its rotation.

The obvious way to completely eliminate this rotation would be to place the center of mass of the *bassbar* under the foot of the bridge, make the support system at the two ends exactly equal and totally decouple the rotation of the *SPF* from the *bassbar*. However, complete elimination is neither necessary nor desirable. It is enough to control the coupling between the *SPF* and the *bassbar*. To this end, we note that the amplitude of the rotational displacement of the *SPF* increases as we move farther away from the *soundpost*, which acts as a fulcrum for rotation. Hence the region between the *SPF* and the *bassbar* appreciably away from the *soundpost* above and below become critical. Fry designates these areas as Stradivari Holes #2, #3 and #4 as shown in Fig. 2. It is the Stradivari hole #4 that is critical in the decoupling of the SPF from the low frequency rotation of the bassbar. Stradivari holes #2 and #3 are important for the high frequency rotational motion of the central part of the bassbar.

Graduations in a Stradivari instrument show thinned areas near the *bassbar* in exactly these regions, implying less stiffness and hence reduced coupling between the *SPF* and the *bassbar* (Fig. 5). The graduations show a variation in thickness from 2.8 mm near the *soundpost* down to 2.1 mm near the *bassbar*. These variations may appear minute, but if we recall, stiffness is proportional to the cube of the thickness, it is a factor of 2.35, a significant enough variation to have a substantial effect. "*You can demonstrate this easily,*" Fry says. "*If you take a violin that has no such thinned Stradivari Holes #2, #3, and #4 and start scraping in these areas, the violin sounds brighter and the shrill region is suppressed.*"

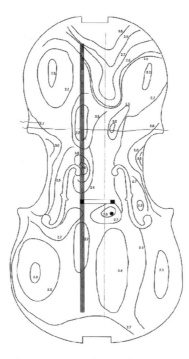

Figure 5: Graduations in a Stradivari instrument showing thinned areas near the bassbar causing the decoupling of the soundpost fibers from the low frequency rotation of the bassbar.

One could have thought of bringing about these changes in the couplings of the bassbar and the SPF by suitably modifying the rotational properties of the bassbar itself by changing its shape and thickness. However, bassbar is frequently adjusted or replaced for other reasons. It is more desirable to install these changes in the plate itself.

Above the shrill region, there is the moderately high frequency region of approximately 3000 to 5000 Hz. Then there is the very high frequency region beyond 5000 Hz. Neither the piston-like action nor the rocking motion of the top can radiate sound in these frequency ranges. Simple physics demands that there be small isolated areas that act as resonating systems, driven and isolated from the rest of the violin body. These are the ***bassbar tweeter*** to the left of the left f-hole and the ***SPF tweeter*** to the right of the right f-hole (Figs. 6–7). They are driven by the resonances in the ***bassbar*** and the ***SPF***, respectively.

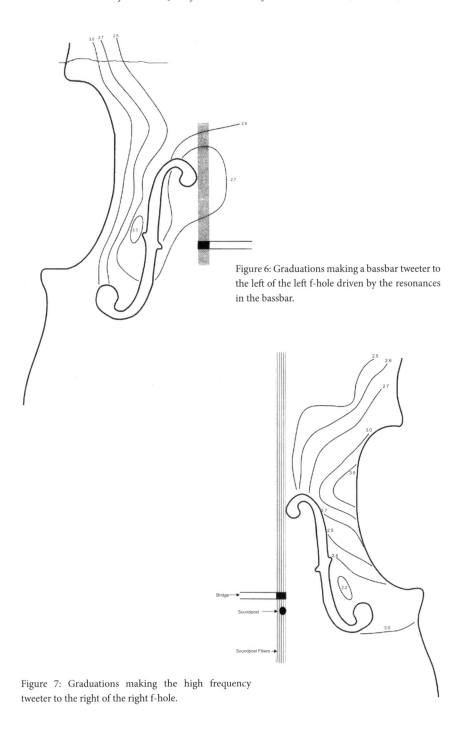

Figure 6: Graduations making a bassbar tweeter to the left of the left f-hole driven by the resonances in the bassbar.

Figure 7: Graduations making the high frequency tweeter to the right of the right f-hole.

The high frequency rotational motion of the ***bassbar*** proved to be a fundamental discovery for Fry. For many years he had thought of the ***bassbar*** as a solid I-beam with displacements and rotations as its primary motions in the low and shrill frequency regions. These motions are governed by its inertial properties, namely, the mass and the moment of inertia about the center of mass. In order to test the correctness of his ideas, Fry conducted a series of experiments in which he attempted to change the moment of inertia by scraping to reduce the mass at various points on the ***bassbar*** (sketched in Fig. 8). When he thinned at the ends in regions around points A and F, the rotational motion was augmented due to the decrease in the moment of inertia, and consequently, as expected, he found an increase in the shrill region. He also expected that scraping closer to the center of mass should have negligible effect, since the decrease in mass would not affect the moment of inertia significantly. However, to his great surprise, as he scraped towards the center, he found significant change in the high frequency range above the shrill region. The closer he scraped towards the center, the more dominant the effect became. One could distinctly hear changes in the loudness of frequencies in the range of 4000 Hz! This surprising discovery clearly needed an explanation.

It then occurred to Fry that he had overlooked a crucial fact, namely, the elastic properties of the ***bassbar***. Any solid transmits sound due to compression waves, the velocity being dependent on the Young's modulus and the mass density of the medium. For spruce, the wood used for the top plate, the velocity is greater than ten times the velocity in air, close to 4000 meters/ second. It was not hard to contemplate the possibility of compression waves, both longitudinal and transverse, set into motion due to the oscillatory force exerted by the bridge via the strings. By using basic arguments of physics, one can show that the longitudinal waves that propagate back and forth with the speed of 4000 m/s can only generate frequencies far beyond the audible range. The case with the transverse waves, however, is different. The ***bassbar***, whose central thick region tapers at the ends, presents a dispersive medium with velocities changing with the thickness. They are reflected at every point as they propagate. This phenomenon coupled with the fact that the center of mass is

Exhibit A

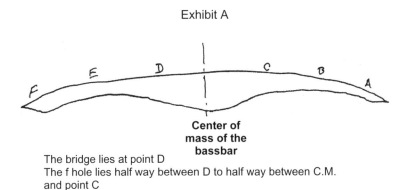

Center of mass of the bassbar

The bridge lies at point D
The f hole lies half way between D to half way between C.M. and point C

Exhibit B
Simple Model of B.B. Refractive

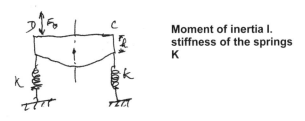

Moment of inertia I. stiffness of the springs K

F_B is the oscillator force which drives this segment of the bridge in a rotational motion

Figure 8: Bassbar resonance system.

displaced away from the foot of the bridge gives rise to internal rotations in the *bassbar*. These rotations flex the *bassbar* and can give rise to resonances. Fry likes to think of this phenomenon in terms of a simple model shown in Fig. 8, a plate of varying thickness with its two ends attached to springs. The spring constants are determined by the stiffness at the ends, and the frequencies of the bending stress waves in the system can be calculated and found to be between 4000–5000 Hz. This was exactly the range of frequencies he heard when he scraped near the center of the **bassbar**. These resonances are coupled to a small area to the left of the left f-hole, giving rise to the **bassbar tweeter** that radiates sound in this mid-frequency range. The graduations shown in Fig. 6 originate from the considerations of making this tweeter system an efficient radiator by isolating it and by adjusting its coupling (impedance) to the **bassbar**.

Finally, the high frequency **tweeter** occupying a small area of a few square centimeters on the right of the right-f-hole is isolated from the rest of the box, except for its coupling to the **soundpost fibers (SPF)** directly over the f-hole. Because of this coupling, the motion of the **SPF** drives the **tweeter**. The graduations in Fig. 7 show thicker contour lines towards the edge. Their thickness varies from 2.7 mm down to 2.2 mm starting from the top of the f-hole and going down. The graduations are such as to make this tweeter radiate frequencies from 10 to 12 KHz down to 6 to 8 KHz. This can easily be demonstrated through a simple experiment consisting of a piece of wood carved to similar dimensions and graduated accordingly. When it is clamped at the thicker end and excited at the other, it indeed becomes a broad band resonator, radiating frequencies of the correct range as specified above.

7

Myth & Reality of Cremona Violins; Fry's Violins

The goal of my research is to produce instruments that have the quality of the great Italian masters, to duplicate the sound of an Amati, a Stradivari, a Guarneri, or a Bergonzi.

William F. Fry

For the last three hundred years, well known luthiers have attempted to replicate the Cremona violins. Although some of them (for instance, Jean Baptiste Vuillaume) have made excellent copies, the general consensus is that they do not come close to reproducing the distinct voices, carrying power, and responsiveness of the instruments of the old masters. Many legitimate scientific researchers have also attempted to demystify the Cremona instruments. Although the studies of the separate components, or "mechanical subsystems," of the violin have provided valuable knowledge regarding how a violin works, they have failed to give any clues about what makes a particular violin stand out among others, let alone revealing the secrets of a Stradivari violin. For three centuries none has been able to do what Stradivari did, to build an instrument that reproduces the range, the power, and expressiveness of the Stradivari sound.

However, the reality is that few genuine 16th century stringed instruments exist. There is only a handful from before 1550. The majority of Cremona violins that are in use have undergone significant changes, dictated by two developments beginning in the early 19th century in the musical scene: the acceptance of a higher concert pitch and the demand for greater volume and brilliance due to increased number of concert goers and larger concert halls. Violin makers found ways to get more sound from the violin by changing the dynamics of the neck, strings, and the bridge to give not only increased power, but also greater bowing facility.

The main stages of changes include raising and arching of the bridge with the consequent lifting of the end of the finger board and hence the neck thrown back at an angle, and a lengthening of the neck by nearly half-an-inch. The higher pitch, increased length and thinner strings of higher density, all caused an increase of tension in the strings. This coupled with the increased angle led to a major increase in the static force on the top plate and required a longer and a more massive bassbar to distribute the resulting pressure over the entire plate. As far as other dimensions are concerned, they have remained standardized since the days of Stradivari. In practically all violins made before 1800 and are in use today, one finds the evidence of these changes.

Nonetheless, in the process of these changes, several characteristics of the originals have survived. The original head and scroll, for instance, have been retained and one can still recognize the works of old masters from their individual craftsmanship, shape and the way the f-holes are cut, and the purflings are inlaid. *"Genuine "strads" have a very characteristic appearance"* says Joseph V. Reid, *"[they] stand supreme in bold and streamlined execution of work. The Stradivari violins display masterfully cut f-holes, exquisitely designed and carved scrolls and harmony of curves, with the entire work set off by a matchless velvety and transparent varnish."*

However, such statements can be very subjective based on historical research conducted by dealers, restorers, and collectors with obviously vested interests. Also because of their exorbitant prices and prestige, one often encounters cases of fraud and forgeries.

Still, according to Fry, in spite of the modernization, a large number of Cremona instruments have retained their distinct acoustical characteristics when compared with instruments made elsewhere. One can compare their tonal qualities with a few that do exist unchanged, the two Amatis and some lesser known ones, that he has played. *"...the interesting thing is they are an outstanding example of the Baroque form, which confirms the fact that the superiority existed in the original form. The acoustics are such that the modernization hasn't destroyed that kind of superiority,"* says Fry. There is undoubtedly a characteristic "Italian" tonal quality in a majority of violins made by the Cremona violin makers in spite of the different methods of construction, differences in arching, etc. Indeed as Vigdorchik notes, *"the arching of the Amati type, rather high, if not steep, at the center, descends gracefully in a slightly scalloped effect toward the edges. At the same time, there is a pronounced thickening of the graduations at the center, while the area near the edges is rather thin."* However, Stradivari, after following Amati until 1700, in his so-called Golden period, made instruments using a completely different method of arching construction and thickness distribution. *"The arching is considerably lower so that the instruments are flatter, with a smoother and more gradual transition toward the edges,"* observes Vigdorchik.

Over the years, in my conversations with several noted soloists playing in symphonic concerts or in chamber music ensembles, I have found invariably their preference for old Cremona instruments, some very well-known, some lesser known. Joshua Bell, Midori and Jennifer Frautschi, who all prefer and play on one or the other Cremona violins of old masters, Stradivari or Guarneri del Jesù. They invariably speak of the superiority of the Cremona violins in the acoustical absolutes over the modern instruments — carrying power, how they sound under the year, how they respond, etc. *"The Cremona tone is a phrase much bandied about,"* says Wifred Saunders, *"There are many people who do not believe in it, but I think we have to presume the players know best and that acoustically these instruments are supreme."*

Jennifer Frautschi, for instance, plays on a Stradivari instrument, made in 1722, named Cadiz. It was owned by Fuchs, a noted violinist who passed away some fifty years ago. *"Some sixty years ago, it went through a major restoration at the hands of John Becker in Chicago,"* says Frautschi. According to Becker, it has undergone the changes in bassbar and neck, but it is one of the most under restored instrument. Frautschi has played on other Strads as well as some modern instruments. *"Modern instruments have the necessary power, but lack in dynamic range,"* she says. *"They are unidimensional with one or two good features. As a player, I find Italian instruments have a response that makes it easy to play, whereas with modern instruments, it is difficult, have to work hard to get the desired notes."*

From the point of view of physics, Fry notes that many of the acoustical properties as a whole are not affected by the mere changes in the shape and size of the bassbar or the neck. They depend upon how different modes are coupled and how they change from one mode to another in rapid succession depending upon the notes played. And these properties are more critically controlled by the thickness graduations of the plates rather than the details of the bassbar. Therefore, it is not surprising, according to Fry, that the old instruments have retained their distinct, tonal qualities and those who listen with discerning ears, recognize immediately the 'voices' of the old masters. We read, for instance, in *Antonio Stradivari*, *"In the hands of great artists, Stradivari violins or violoncellos will entrance by their quality, astonish by their force and depth, and stir all the emotions by their varied capacity for expression. And these instruments, produced two centuries ago, are still for all practical purposes superior to any which have since been made."*

Over the years, Fry has studied, designing numerous experiments and testing the changes in the acoustical properties brought about by varnish, asymmetric graduations, changes in bassbar shapes and sizes and graduations in localized critical areas described in the earlier chapters. *There is no one or two single 'secrets' that make a Stradivari or a Guarneri or a good violin in general.* Varnish with its polymer properties changes the stiffness in the direction transverse to the grain in the wood and stiffens the wood. From unvarnished violins, made in German factories, Fry could make better sounding instruments with increased volume by controlling the penetration depth, the number of coatings, the drying temperature, etc. Introduction of asymmetric graduations around the soundpost in the back plate enhanced the "breathing mode" and made instruments that had greater depth in the low frequencies. With numerous experiments, Fry gained greater and greater understanding of the complex interplay of the various changes and their effects on the tonal qualities of the instruments. From old 19th and 20th century copies of Cremona violins he acquired in antique shops and violin stores, he made some fine sounding instruments. Some professional players and young upcoming violinists have bought them at affordable prices and have played on them at concerts.

During the past two years, Fry has brought together all his ideas and focused his attention on his ultimate quest: *to duplicate the subtle tonal qualities attributed to a Stradivari.* He aimed to create an instrument to demonstrate, an instrument endowed with all the acoustical absolutes — a wide dynamic range, *quick and easy in response for a player, soft under his ear, but radiating a "soprano" like silky voice for the listener.*

According to Fry, the following aspects are crucial in reproducing the sound of a Stradivari instrument:

- Each bowed note creates the fundamental accompanied by its overtones that encompass a range of frequencies that include shrill as well as high frequencies with their refined silky quality. It is extremely important that there be a proper balance of the harmonic content with enough high frequency content to provide the carrying power, enough suppression of the shrill components and enough of the low frequency components to realize the sounding of the fundamental.

- Ringing, especially High Frequency Component ringing.

These features depend on the graduations in the plates, most critically on the graduations in the top plate, which incorporates the high frequency tweeter.

The starting point for Fry is the choice of a good copy of a Cremona instrument, modeled after a Stradivari in its contour, shape, size, arching and the shape of the f-holes. It is well known that Stradivari used different blocks to shape his instruments during different periods. During his early period, he followed Nicola Amati, but soon started making gradual changes, working the scroll in more boldly and in greater detail. He modified his pattern, narrowing its shape in the C-bout, elongating the instrument and changing the shape of the arching. *"If I had infinite possibilities,"* says Fry, *"I would choose the period in which I find a Stradivari copy that I like best. In general most people like the later period instruments better than the earlier ones, although the earlier ones have their own wonderful qualities."*

Fry prefers to work on copies of instruments of Stradivari made during the period known as the Golden Period (1700–1720). It is probably the most distinctive because of the 'tonal color and silky brightness' of the sound of the instruments made during this period. The instruments made during this period are also the most copied ones. He has several copies for his research and a couple for final modeling. Even though such copies may have the correct outline and shape, details of the arching are subtle and difficult to copy. *"I don't find in such copies arching exactly the way I like,"* Fry says. *"I have to live with that. And wood, of course. No two pieces of wood are the same. Ultimately the final graduations have to compensate for the variations of woods, shape, arching and most of all, varnish. Varnish changes the stiffness properties of the wood, changes that depend upon so many factors. So I get a box. I don't know its history of varnishing. I don't know the properties of the wood. I have to do the final control by scraping on the inside after it's all done."*

Fortunately for Fry, most violin makers through the centuries have made the violin plates way too thick. The reason, according to Fry, is that a violin with thin plates graduated incorrectly sounds 'horrible,' whereas a violin with thicker plates may not have that 'gorgeous' sound, but it responds better and it is playable. Rather than risk with thin plates with incorrect graduations, they play it safe by keeping the plates thick. This allows Fry to alter the thickness graduations as desired without resorting to inlays in many cases. Inlays present difficult problems, since the elimination of the stiffness of the glue requires careful placing of the inlay surface at the right depth in the plate.

After selecting a good model with varnish that appears reasonable, Fry says *"I take it apart, take the top of, remove the bassbar. The first thing I do is make a map of the thickness graduations of the top and back plates. If I am fortunate enough, there is plenty of wood everywhere and then the problem is just one of taking the wood away. I have to decide what kind of instrument I want to make, soprano or deep sounding violin?"* From numerous experiments and Wilson Powell's measurements, Fry has a very good idea regarding the maximum thickness needed in and around the critical areas (see Fig. 2 in Chapter 6). For a soprano instrument, for instance, the central region in the back plate needs to have a thickness ~4.5 mm. On the other hand, for an instrument to be deep sounding, it has to be thinner with ~3.9 mm. These differences may appear minute and inconsequential, but they are indeed significant when we recall that stiffness is proportional to the third power of the thickness.

The next step is to build proper asymmetries around the sound post to enhance the piston-like motion and the breathing mode. They also depend upon what type of instrument one wants at the end. From the point of view of the tonal quality and carrying power, it is the top that plays singularly significant role compared with the back plate. Consequently, thickness graduations in the top plate are most crucial. So is also the shape and size of the bassbar with its high frequency resonances. Fry often finds it necessary to replace the bassbar with one of his own designs.

For Fry, a focused, 'soprano' like silky voice of an instrument is crucially dependent upon the radiation of the highest frequency components combined with the *high frequency component ringing (HFCR)*. Now, when the fundamental is played on a violin string, it generates all its overtones. For the tonal quality of a Stradivari that Fry is after, it is specifically, the *HFCR* between different strings that is crucial. By this what he means is the *exact matching* of the high overtones of the played string with overtones of an open string. When this happens, there is an energy transfer from the bowed string to the open string, as well as energy loss to the box. As a consequence, the timber of the played note undergoes a change.

Why this is so important? If there is exact matching of the high overtones between different strings, then in normal playing, a slight oscillation of the finger with a barely perceptible variation in frequency, is sufficient to produce perceptible changes in the timbre. This is particularly important when playing a *vibrato* for instance, since the changes in the timbre due to ringing enhances the perception of *vibrato*. Moreover, this kind of change in timbre is instantly perceived by the player and assures him that he is playing in tune. It should be noted, however, the whole phenomenon of high frequency ringing is dependent upon the violin radiating the high frequencies.

The art and science of Fry in achieving this type of ringing and thereby enhancing the refined silkiness of the voice in the radiated sound, and at the same time retaining a soft, shrillness suppressed local sound, consists in designing a complex high frequency filter between the *Soundpost Fibers* (SPF) and the high frequency **tweeter**. As Fry says, SPF *"are the nerves and muscular system"* of the violin. The physical vibrations of the right-foot bridge set the SPF into vibrations and rotations. The efficient radiation of the high frequencies depends upon their coupling to the tweeter just above the top of the right f-hole. At the same time, the suppression of the shrillness depends upon their rotational properties in the lower bouts.

To achieve these dual ends, Fry notes the crucial role of two elements of the top plate: (1) the neighboring fibers to the right of the **SPF** that he designates as *Transfer Fibers* (**TF**); (2) The small region, the *engine*, surrounding the *Soundpost* (**SP**) and the bridge right-foot. This is the place from where all the driving forces originate and create the rotational and vibrational modes of the **SPF** (Fig. 1). The **TF** to the right couple the **SPF** to the **tweeter**. In order that these intervening **TF** provide faithful filter communicating the high frequency radiation from the **SPF** to the **tweeter**, they have to follow faithfully the motion of the **SPF**. Now the **SPF** flex and rotate around the **SP** acting as the fulcrum (nodal surface). In order that the **TF** to the right follow this motion, they require a nodal point. Fry creates such a nodal point by thickening (adding mass to) a small region below the **SP** and to its right that he designates as M_1 (Fig. 1). Secondly, the efficient coupling of the right side of the **TF** to the high frequency **tweeter** depends upon the relative acoustical impedances between the two. The tweeter has relatively high impedance (ratio of the applied force to initiate a given velocity) whereas **TF** have relatively low impedance. To overcome this imbalance and make the **TF** transmit effectively the very high frequency components to the tweeter, one needs to create a lever system between them, which effectively increases the coupling of the TF to the tweeter.

The obvious solution that suggests itself is to stiffen the fibers by increasing thickness. Increasing thickness, however, results in the increase in mass. Consequently, the increased inertia reduces the very high frequency motion of the **SPF** and the **TF**. After several experiments, Fry convinced himself the solution consisted in strengthening the coupling of the motion of the bridge right-foot in the region marked as the **engine**.

However, in doing so, one has to contend with the fact that spruce (or that matter any wood) is strong along the grain and weak in the transverse direction. The ratio of Young's modulus parallel and perpendicular to the fibers varies from 12–15 depending upon the tree and the wood. Fry needed to

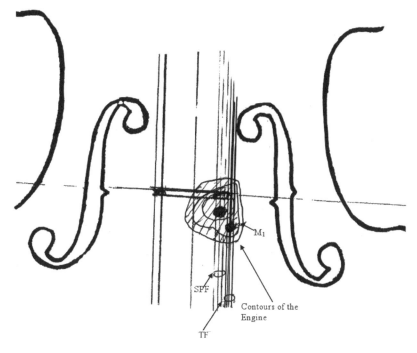

Figure 1: The engine, soundpost fibers (SPF) and transfer fibers (TF).

change this ratio to strengthen the coupling without adding mass, that is, *"to put something in the wood that increased the stiffness in the transverse direction without changing significantly in the direction parallel to the fibers."* Again, an obvious solution is to use some varnish. But the problem with varnish is it takes some two years or more for it to reach its final stable condition. That clearly was not suited to Fry's mode of research.

He came up with an ingenious solution of using normal Elmer's wood glue. With proper dilution with water (1/2 glue, 1/2 water) to reduce the viscosity and soak into the wood better, he found glue an ideal substitute for varnish. It has the same properties as varnish; it strengthens in the transverse direction and has the added advantage that it dries up and reaches a stable condition in a matter of few hours. He quickly found the idea worked; it had the anticipated effect on the tonal quality of the violin.

With the possibility of controlling the stiffness in the transverse direction of the fibers, Fry experimented with the effect of changes on the tonal qualities by changing other elements to the right and left of the **SPF**. He had new ideas about the length, shape and thickness graduations of the **tongue** with the Stradivari model in mind. The design of the **tongue** (see Fig. 4 in Chapter 5) is such that it effectively acts like a lever between the **bassbar** and the **SPF**. The **tongue**, where it is close to the **bassbar** follows the latter's motion, magnifies it and communicates it to the **SPF** at the other end so that the **bassbar** and the **SPF** move in phase. An equally important effect of the **tongue** is to shorten the length of the **SPF**, reducing their effective length essentially by the length of the tongue. Since the **SPF** are very flexible compared to the **bassbar**, this shortening stiffens them around the vicinity of the **soundpost** and this in turn reduces the tendency of rotation around the soundpost. The double bound fibers (Fig. 2) also play an important role in the final tone of the violin.

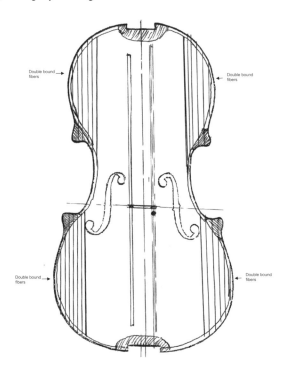

Figure 2. Double bound fibers.

This brings up a very intriguing speculation on the part of Fry, namely, whether the Cremona violin makers did indeed use this idea on a finished, but as yet to be varnished instrument. *"If you applied diluted transparent varnish to the small region, you can put as many coats as you want,"* says Fry. *"The varnish will impregnate the wood in that small region. I had this sort of intriguing idea that the Italian makers could have used uncolored varnish and impregnated that little area by applying it many times before letting it dry. Then put the normal varnish with color that covers it up. One would never be able to tell. A bold speculation, but that would be one way one could do that. It could possibly be one of Stradivari's secret."*

The most important, perhaps most radical innovation of Fry in his quest for reproducing the sound of a Stradivari, is making predictable changes in the sound after the violin is made. The graduations made in the plates and the associated normal modes change significantly in the completed varnished body of the violin. Using simple scraping devices, Fry can change the timbre to achieve a desired tonal quality. Fry calls it "fine tuning." In judging the changes, it is important to recognize the qualitative difference between the timbre of the local sound the player hears and the radiated sound the listener responds to. In reproducing a desired tonal quality, it is the latter that is most important. *"So I can hear the radiated sound rather than the local sound, I use a good recording system and record the results on a CD from a microphone at a distance of several feet. I can hear much more and learn more from the recorded sound than just playing the violin."*

In the video accompanying the book, Fry with the help of Rosemary Harbison, demonstrates his fine tuning process. He shows how minute changes in critical areas of the top plate bring about predictable changes in the tonal qualities of an instrument. The violin he uses is a copy of a Stradivari instrument in the late period made in France in 1920. It has a relatively low arching, but he has built in all the normal features of a Stradivari model — tongue, tweeter, bassbar resonance and Stradivari holes 2 and 4. The central region of the back plate, however, is thicker than normal with 4.8 mm.

Initially the sound of the violin is metallic, shrill and loud under the chin. It has limited dynamic range and does not have the desired high frequency ringing. Fry specifies the changes needed to overcome these short comings. As one observes and listens, one wonders at the changes in the tonal qualities by minute, miniscule changes in thickness graduations in critical areas. At the end of the demonstration, we have an instrument substantially improved with reduced loudness under the chin and with improved low pressure response. One perceives a more focused sound with improved high frequency ringing. However, it remains with an exaggerated soprano voice because of the excessive thickness in the central region of the back plate.

In the final moments of the DVD, Harbison plays on an instrument that Fry has rebuilt with the objective of achieving his goal to reproduce the tonal qualities of a Stradivari instrument. The instrument he started with was also a copy of a late period Stradivari instrument with a label indicating that it was made in Germany around 1932. In its original form, the measured thickness of the central regions of the back and top plates exceeded the normal (back plate 5.8 mm, top plate 3.8 mm) and did not have any of the features that Fry considers vital for a good sounding instrument. With appropriate reductions, carving a tongue and constructing the tweeter, bassbar and introducing all other features discussed earlier, Fry arrived at the thickness graduations for the top and back plates shown in Figs. 3 and 4.

The segment Harbison plays is the slow movement of Brahms violin concerto. The violin has an excellent quick response with a clarity in the fast passages and has the sought-after high frequency ringing, which is quite

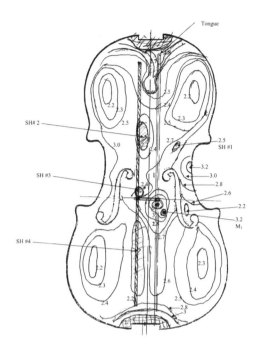

Figure 3: The thickness graduations of the top plate of the reconstructed Stradivari model of Fry.

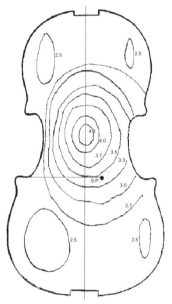

Figure 4: The thickness graduations of back plate of the reconstructed Stradivari I model of Fry.

evident in the vibrato. It has a high frequency silky radiation which contributes to its beautiful sound and also to its carrying power. Most people would say the sound is very colored. And Harbison comments that it has a large dynamic range, easy to play.

As Fry comments at the end, his rebuilt instrument has all the acoustical qualities of a Stradivari instrument that he was aspiring to achieve. *"Yet, it may not be perfect in the sense that it may be what A likes, but not B,"* Fry says, *"It is a matter of aesthetics now."* It is not different from the opinions of the players and musicians who speak of one Stradivari instrument superior/inferior compared to another. No two Stradivari instruments sound alike. At this point we can say we transcend the realm of objectivity and enter the realm of intangibilities dominated by mythology, sociology and cultural history.

8

A Convergence of Science and Art

...the core principle of new experimental science was that what one person can discover any other can find again — yet for three centuries none have been able to do what Stradivari did, to build an instrument that reproduces the range, power, and expressiveness of the Stradivari sound.

Thomas Levinson

Fry recalls an incident when he was giving a colloquium in Edmonton, Canada on Cremona violins and violin acoustics. As he was talking, he couldn't help but notice a middle-aged man sitting in the front row, visibly angry, nervous and upset at what he was hearing throughout the talk. When Fry had finished his talk, the man stood up, turned around, faced the audience and said, *"You know Professor Fry, your problem is you don't know how to put the soul in the violin. The Italians learned how to do that. And that is your problem. Until you learn how to put the soul into the violin you will never make a good sounding violin,"* and started walking out. Fry was stunned for a moment, but as the gentleman was exiting, Fry responded, *"Well, I couldn't agree with you more. The only difference is I'm trying to define the soul you are talking about."* The man turned around, looked at Fry and dashed out.

For Fry, the soul of a violin is *simple physics.* And contrary to Levinson's assertion, Fry's demonstrations and his success in reproducing the tonal qualities that Levinson ascribes to a Stradivari instrument mark a historical achievement in violin acoustics. As Fry says, the violin presents a very complex instrument acoustically. The conventional scientific approach to reduce it to an analysis of its subcomponents and to try to find one or two "secrets" such as varnish or special qualities in the wood have failed to reproduce the desired acoustical features of the old Cremona instruments. When the celebrated violinist Joshua Bell says, *"he prefers a 1713 Gibson ex Huberman Stradivarius with its good balance between the sweetness, which Strands tend to have, a sort of soprano, feminine sweetness to them, but also, has a very powerful lower register, it's very well-balanced, and it projects well in the hall,"* he is describing his subjective experience in qualitative terms, none of which can be given a precise quantitative characteristic that is amenable to a physical measurement.

Fry's holistic approach based on sound principles of physics ushers a new era in violin making. It departs from the main focus of the traditional violin makers on producing copies of old Cremona instruments, which involves mainly copying the geometrical dimensions, making accurate arching templates, working with precisely copied modeling boards and so on. However, mere "mechanical" copies could never equal that of the original instrument in the tonal qualities of the instrument. Even if the copying is perfect, the variations in the acoustical properties of the wood and varnish would never reproduce the tonal properties of the original instrument. Besides, as Martin Schleske says, "*The golden age of classic violinmaking was actually a product of a living tradition that came to life over generations. Any attempt to "conserve" this tradition would necessarily break with it. Anything that improved the sound was kept, and everything else was discarded. Each generation passed its experience to the next in a process of constant trial and error. This meant the heyday of the violin represented the result of centuries of empirical research.*" A modern violinmaker does not have the luxury of that rich inherited experience.

Fry, however, with decades of semi-empirical research working with what he calls "old junk violins" from the nineteenth century, has provided invaluable guidelines in making great violins. Many of his modified and transformed violins, made during the course of his research, have been in the hands of young and upcoming violin players, who use them in concert playing. By isolating critical areas in the top and back plates and demonstrating how minute changes in these areas control the tonal quality and why, Fry has advanced a new approach based on the interplay between art and science. While the main focus in this book has been on reproducing the sound of a Stradivari instrument, Fry's techniques do not preclude reproducing other sounds, the sound of Guarneri instruments, for instance, which are described as more colored, more in the fundamental compared with a Stradivari silky, soprano-like sound with exaggerated high frequencies. Fry continues his research with his personal goal to duplicate in a predictable manner the sound of other old Cremona instruments, an Amati, a Guarneri, or a Bergonzi. The final judgment regarding whether he has finally solved the mysteries of the Cremona violins rests with the violin makers, violin players and discerning listeners.

Concluding, Fry's success, however, must not be construed as implying that there is now a precise mechanical method for turning out great violins, each sounding exactly like another. Craftsmanship, quality of wood, varnish, and attention to other details leave the final result open to wide variations. However, a violinmaker who understands the science of his instrument can make consistently better instruments, perhaps one day even better than those of the old Cremona masters. It is also well to remember, with all due respect for the rich heritage of centuries, not all the old Cremona instruments, valuable as they are, possess the marvelous tone and performance one expects from such masterpieces. William F. (Jack) Fry's effort, to understand and define the "soul" of the great violins, like the best scientific and artistic inquiry, is an unending quest for truth, enlightenment and beauty. Hope it spurs others to carry on this important work into the future.

Notes and References

Chapter 1

Tarisio's story is to be found in several books. The accounts vary slightly in details, but not in the main substance. Principal sources for my account are the following:

1. _The Violin Hunter,_ by William Alexander Silverman (The John Day Company, New York, 1957), a book that reads like a fiction, but according to the author, is based on extensive research. As Silverman writes at the end of his book, "With the exception of some of the dialogues, which I have used to help define the presumed motives of some of the people in it, this story has been reconstructed from historical fact. Even the dialogue, a considerable amount is documented; the rest might rank as reconstructed fact, being based on stories repeated by the principals to the others." Tarisio's encounter with Sister Francesca is found only in this book (pp. 40–47). Here is a colorful account published in _The New York Times,_ June 24, 1883, which supports Silverman's description of Tarisio's acquisition of the old lost Cremona instruments:

TARISIO AND THE CREMONAS:

At the beginning of this century, hidden away in old Italian convents and wayside inns, lay the masterpieces of the Amati, Stradivarius, Guarnerius, and Bergonzi, almost unknown and little valued. But Tarisio's eye was getting cultivated. He was learning to know a fiddle when he saw it. "Your Violino, Signior, requires mending?" says the itinerant peddler, as he salutes some monk or padre known to be connected with the sacristy or choir of Pisa, Florence, Milan. "I can mend it." Out comes the Stradivarius, with a loose bar or a split rib, and sounding abominably. "Dio mio!" says Tarisio, "and all the blessed saints, but your violino is in a bad way. My respected father is prayed to try one that I have, in perfect and beautiful accord and repair and permit me to mend this worn-out machine." And Tarisio, whipping a shining, clean instrument out of his bag, hands it to the monk who eyes it and is for trying it. He tries it: it goes soft and sweet, though not loud and wheezing, like the *battered old Strad. Tarisio clutches his treasure. The next day back comes the peddler to the cloister, is shown up to the padre, whom he finds scraping away on his loan fiddle. "But," he exclaims, "you have lent me a beautiful violino and in perfect order." "Ah! If the father would accept from me a small favor," says the cunning Tarisio. "And what is that?" "To keep the violino that suits him so well, and I will take in exchange the old machine which is worn out, but with my skill I shall still make something out of it." A glass of good wine or a lemonade or black coffee clinches the bargain. Off goes Tarisio, having parted with a characterless German fiddle — sweet and easy — going and "looking nice," worth now about £5 — in perfect order, no doubt — and having secured one of those Gems of Cremona which now run into the £200. Violin-collecting becomes the passion of Tarisio's life.*

The story has been told by Mr. Charles Reade, and all the fiddle world knows how Tarisio came to Paris with a batch of old instruments, and was taken up by Chanot and Vuillaume, through whose hands passed nearly every one of those chefs a 'oevre recovered by Tarisio in his wanderings, which now are so eagerly contended for by English and American millionaires, whenever they get into the market.

2. *Violins and Violinists,* by Franz Farga (Rocklift, Salisbury Square. London, 1940). The quote near the end of the Chapter, what Vuillaume found in the attic, is on p. 101.

3. *Old Violins and Violin Lore,* by Rev. H.R. Haweiss (William Reeves, publisher; 1921).

4. *Count Ignazio Alessandro Cozio di Salube* (Technical Studies in the Arts of Musical Instrument Making, Andrew Dipper and David Woodrow, First Edition. ISBN 1 870952 00 6; Photoset, printed and bound by Redwood Burn Limited, Trowbridge, Wiltshire).

5. The violin, *Le Messie* or *Messiah* rests now at the Asmolean Museum in Oxford, England. Its amazing condition, its youthful appearance and glowing varnish has given rise to speculations regarding its authenticity. It is conjectured that Vuillaume and not Stradivari actually made this instrument at a much later date and propagated the history that it never

had for his own benefit. Recently, however, John Topham, using the science of *dendrochronology*, has made a tree ring analysis of the *Messiah*. He has demonstrated that the ring widths on the top and bottom plates of *Messiah* match quite well with those on another very well established violin made by Stradivari. (John Topham in the *Newsletter of the British Violin Making Society, Issue 13, Autumn 98, pp. 7–8*. Also, see the cover story in The Times, October 27, 2000, *A Stradivarius riddle by* Guiles Whittell).

Picture Credit: The picture of *The Sigismondo Church* with its interior is from Cremona; Art City, Regione Lombardia — Azienda Di Promozione Turistica del Cremonese.

Chapter 2

1. *The Story of the Violin* by Paul Stoeving (The Walter Scott Publishing Co,. Ltd London, New York: Charles Scribner's Sons 1904).
2. *The Physics of Violins* by Carleen Hutchins (Scientific American, November 1962).
3. In attributing the violin's origin to India, F.J. Fetis, says in *"Notice of Anthony Stradivari, The Celebrated Violin-Maker,"* known by the name of, *Stradivarius*, translated by John Bishop, William Reeves, London, 1964:

   "The country which affords us the most ancient memorials of a perfect language, of an advanced civilization, of a philosophy where all directions of human thought find their expression, of a poesy immensely rich in every style, and of a musical art corresponding with the lively sensibility of the people — India, appears to have given birth to bow-instruments, and have made them known to other parts of Asia and afterwards to Europe."

   He describes "Ravanastron" as
   A small hollow cylinder of sycamore wood, open on one side, on the other covered with a boa skin (the latter forming the sound board), is traversed by a long rod of teal — flat on top and rounded underneath — which serves

as neck and fingerboard, and is slightly bent towards the end where the pegs are inserted. Two strings are fastened at the lower end and stretched over a tiny bridge, which rests on the sound board, and is cut sloping on top. A bow made of bamboo — the hair roughly attached on one end with a knot, on the other with rush string — completes the outfit.

(The instrument is exhibited in the ethnographical department at the British Museum. It is among the exhibits from the hill tribes of Eastern Assam.)

Stoeving about Ravanastron:
It is the unmistakable family likeness which links together the old and the new, the crude and the perfect, the Ravanastron and the sovereign Strad.

4. For an account of the origin and its history, see Groves Dictionary of Music, pp. 806–812. See also, *The Quagmires of History* (First Chapter in *Four and Twenty Fiddles* by Peter Holman, Clarendon Press, Oxford 1993).

5. *Some thoughts on the Acoustics of Bowed Stringed Instruments* by Isaac Vigdorchik in the *Journal of the Violin Society of America,* 5 (3) 1979, 26–71. The pitch or the perceived frequency of a tone or note is determined by tapping the plates at precise locations. Generally speaking pitch of a plate is determined by tapping at the center of the plate.

6. *The Violin Makers of the Guarneri Family (1626–1762),* William Henry Hill, Arthur F. Hill and Alfred E. Hill.

7. *La Cremona di Stradivari* (Stradivarian Itineraries), Editrice Turris, Cremona 1988. Record in the Notries of Cremona shows that Stradivari paid 7,000 lire imperiali for the house he bought, 2,000 in cash, and the balance of 4,990 in four years. An allowance of 10 lire was made because of his religiosity; his contribution of 6 lire/ year to the cathedral.

8. *Antonio Stradivari*; Master Luthier. His Life and Instruments by W. Henley (Amati Publishing Ltd., Brighton 1. Sussex, England).

9. *The Violin Hunter* by William Alexander Silverman (John Day Company, New York, 1957).

10. In the preface of the book, *Antonio Stradivari; His Life and Work by* W. Henry Hill, Arthur Hill and Alfred E. Hill (Dover Publications, Inc., 1963), the authors credit Signor Mandelli's researches in the archives of his native city of Cremona. With new information gathered, he approached them in 1890 to publish the material in English and Italian. After their own further efforts, the book was first published in 1902.

Figure and Picture Credits: Figure 1. *La Cremona Di Stradivari (Stradivari Itineraries), Editrice Turris-Cremona-1998.* Figures 2, 3 and 4 are all from *Antonio Stradivari, His Life and Work,* W.H. Hill *et al.*

Chapter 3

1. Principle source, conversations with William F. Fry.
2. *The Glory of Violin* by Joseph Wechberg (The Viking Press, New York, 1973).
3. Groves Dictionary of Music Instruments and musicians.

Figure Credits: Figure 1 reprinted from *Wisconsin Academy Review,* Wisconsin Academy of Sciences, Arts and Letters, Spring 2000, Vol. 46, No. 2. Figure 2 from *The Violin Family,* D.D. Boyden et al. (W.W. Norton & Company, New York, London, 1989). Figures 3 and 4 from *The Cambridge Companion to the Violin,* Ed. R. Stowell (Cambridge University Press 1992).

Chapter 4

1. *On the Sensations of Tone,* Hermann L.F. Helmholtz (Dover Publications, Inc., New York 1954). Helmholtz in the Introduction,

Each organ of sense produces peculiar sensations, which cannot be excited by means of any other; the eye gives sensations of light, the ear sensations of sound, the skin sensations of touch. Even when the same sun beams which excite in the eye sensations of light, impinge on the skin and excite its nerves, they are felt only as heat, not as light. In the same way the vibration of elastic bodies heard by the ear, can also be felt by the skin, but in that case produce only a whirring fluttering sensation, not sound. The sensation of sound is therefore a species of reaction against external stimulus, peculiar to the ear, and excitable in no other organ of the body, and is completely distinct from the sensation of any other sense.

2. The legend of Pythagoras; how he discovered consonance ratios by listening to the sound of hammers striking an anvil, but later used monochord to establish the ratios is described in *Measure for Measure; A Musical History of Science*, Thomas Levenson (Simon & Schuster, New York 1994): 22–24.

3. *Dialogues Concerning Two New Sciences*, Galileo Galilei (Dover Publications. Inc., New York 1954): 103. Young Galileo participated in the experiments and one might say he received from his father the important aspect of the scientific method — testing theoretical ideas by appeal to experiments.

4. *Harmonie Universelle*, Marin Mersenne. His monumental work on music and musical instruments was published in 1636. In 1625, he published his observations on the stretched string and the mathematical law quoted in the text. However, there is some confusion about whose work, Galilei or Mersenne, was reported first, and both Galilei and Mersenne are given credit.

5. Some other notable developments during this period concern the Frenchman Joseph Sauveur, who emphasized that harmonics (overtones with frequencies that are integral multiples of the fundamental frequency), are components of all musical sounds. He was a mathematician who was drawn to the subject of 'acoustics' (so named by him) through an interest in music. He explained how a string can vibrate at higher frequencies by dividing itself into multiple segments of equal lengths separated by stationary points (nodes, *noeuds* in French) and the points of maximum motion (anti-nodes, *ventres* in French). Fundamental had no nodes; it had only one segment.

Pierre-Louis Moreau de Maupertuis (1698–1759), also the renowned physicist from France, noted for his "principle of least action," advanced a theory in 1726 in which different tones in the violin were carried by the so-called "reeds" (fibers) in the spruce wood of the top plate. He believed that the different lengths of the reeds, controlled by the shape of the top plate and the cutting of the f-holes, affected the sound in different tone ranges (*Musical Overtones and the Origin of Vibration Theory*, Sigalia Dostrovsky, CAS Journal, Vol. 4, No. 1 (Series II) May 2000; see also *350 Years of Violin Research: Violin Development from the 16th through the 19th Centuries*; Research Papers in Violin Acoustics 1975–1993, Ed. Carleen Maley Hutchins).

6. Chladni's studies of the vibrations of membranes and plates and his popular expositions with scientific demonstration experiments at the time of French Revolution had a great impact on physicists and mathematicians (*Acoustics for Violinmakers: Theory and Practice*, Anne Houssay (unpublished)). It also led to experimentation with violinmaking. In 1800, John J. Hawkins of London sacrificed a Strad to make a violin without ribs and back; he was granted a patent. Various other attempts changing the sizes and shapes of violins followed. But Franz Anton Ernst, a concert master and a violinmaker, declared in an article published in *Allgemeine Musikalische Zeitung*, that after twenty years of experimentation he had reached the conclusion that nothing about the old masters' called for improvements, especially their shape.
(*The Amadeus Book of the Violin*; Construction, History and Music, Walter Kolneder, Translated and edited by Reinhard G. Pauly, (Amadeus Press, Portland, Oregon 1998): 181)

7. A report about Chanot's success was published in *Moniteur Universel*, of 22 August 1817, stressing how surprising it was that Chanot's violin compared successfully with the Strad, "*even though it had been made from wood cut only six months ago.*" (ibid. *Amadeus Book*)

8. *Some Thoughts on the Acoustics of Bowed and Stringed Instruments*, Isaac Vigdorchik, Journal of the Violin Society of America, 1975, 5, 26–71.

9. J. Woodhouse and P.M. Galluzzo, *The Bowed String as We Know it Today*, Acta Acustica United with Acustica Vol. 90 (2004) 579–589 provides an excellent review of early work beginning with Helmholtz leading to the 20th century research.

10. Main source for 20th century research, *Research Papers in Violin Acoustics 1975–1993*.

11. *Journey into Light; Life and Science of C.V. Raman*, G. Venkataraman (Penguin Books India 1994).

12. Carleen M. Hutchins, *Acoustics of Violin Plates, Scientific American* Oct. 1981, pp. 170–186. Free plate tuning, however, has its limitations in that the tuned modes of free plates disappear into the body modes, when the instrument is assembled and varnished. See Joseph Curtin, *Tap Tones: An Update*, CAS Journal, Vol. 5, No. 1 (Series II) May 2004; pp. 8–10.

13. G. Bissinger, *Modern Vibration Measurement Techniques for Bowed String Instruments, Feature Article in Testing of Acoustic Stringed Musical Instruments: Part 4, July/August 2001* EXPERIMENTAL TECHNIQUES: 43–46.

14. Martin Schleske, *Wired for Sound, The Strad October 2002*: pp. 1111–1114. Also *Tonal Copies in the Field of Violin Making*, (Meisteratelier für Geigenbau München): pp. 1–9.

Chapter 5

1. The principal source for this chapter is extended taped conversations with Fry. Mrs. Audrey Fry, who went to grade and high school with Fry, contributed to the biographical account of Fry's earlier years.

2. Among many stories about the special nature of wood accounting for the exceptional quality of Stradivari sound, is one that received great publicity in December 2003. Lloyd Burckle, a paleo-climatologist at Columbia University and Henry Grissino-Mayor (BGM), a dendrochronologist at the University of Tennessee, claimed there was a distinct narrowing of rings in the period 1620 to 1715 AD. This narrowing ring pattern appeared to coincide with the middle of a long climate cooling period mid 1400s to the mid 1800s known as the Little Ice Age, but also very closely with a period of time 1645–1715 when sunspot activity was at a distinct minimum known as the Maunder Minimum (after its discoverer E.W. Maunder, a 19th Century solar astronomer). BGM surmised *".. trees were under stress and were only able to lay down narrow rings... the acoustical qualities of this narrow grained wood may be superior to the wider grained wood used by makers before and after. This gave Stradivari a distinct tone preferred by people then and now."*

   John Topham in the article, *Stradivari and the "Little Ice Age"* (News Letter of British Violin Making Association (BVMA) Issue 35, Spring 2004) characterizes this as *"idea flawed, far-fetched and not supported by any direct evidence."* The whole paper, according to him, was a speculation based admittedly on professional insight into climate but on no actual facts relating to the instruments at all. He goes on to say

   "there is nothing to suggest Stradivari wood is either particularly narrow grained or appears to have any other special quality about it. From about 90 instruments I have seen, dating between 1666 and 1736, the wood on the fronts is very similar in its ring width variations to the wood used not only by his contemporaries, i.e. Guarneri del Gesu, Guadagnini and the others, but also by makers before and after, even up to the present day."

3. Polymers in general have the property of forming large molecules with repeating structural units typically connected by covalent chemical bonds. Cellulose, the main constituent of wood and paper is a natural polymer. When varnish polymerizes, the cellular structure alters the elastic properties of the wood by strengthening the strength of the fibers in the direction perpendicular to their length.

4. Stiffness is characterized by what is called Young's modulus, which measures the ratio of the applied stress to the strain of a material.

5. *Antonio Stradivari: His Life and Times* by W. Henry, Arthur F and Alfred E. Hill (Dover Publications., New York 1963) pp. 184–86.

6. Savart had access to these instruments through Vuillaume (*Memoire sur la construction des instruments á corde et á archet*, 1819).

7. For Instance, a Stradivari owned and played by the late Leonard Sorkin during his early years had no flames at all on the back. It was as plain as the board on a kitchen sink, says Fry.

8. Kim Won-Mo, a renowned concert violinist was also an early participant in Fry's research and used some of Fry's violins in concerts.

Chapter 6

1. Indeed, among the exhibits in the Stradivari Museum in Cremona, one finds the tools Stradivari used. Among cuttings and wood pieces that were intended to be made into violins, there is a wedge-shaped tool to measure the thickness and also a scraping device. During a visit to the museum, Fry and I discovered with great excitement the scraping device with an uncanny resemblance to Fry's "scrapers." It vindicated his theory that Stradivari did make final adjustments by scraping and changing the graduations through f-holes in a manner like his own method to change the tone of a violin.

2. In Stradivari instruments, maximum thickness occurs almost at the geometrical center, and hence displaced to the left and above the *SP*. In the case of Guarneri instruments, it is closer to the *SP*, resulting in greater depth of the sound in lower frequencies.

3. Moment of Inertia is what determines the rotational motion of an object about an axis. It depends upon the mass and its distance from the axis. Conceptually, it is analogous to mass in translational motion.

4. Sound waves in media are longitudinal waves. The propagation of the wave is in the same direction as alternate compressions and rarefactions of the medium that transmits the wave.

Chapter 7

1. *YOU CAN MAKE A STRADIVARI VIOLIN*, by Joseph V. Reid (Williams Lewis & Son, 1967, Illinois, U.S.A).

2. Wilfred Saunders, in the Proceedings of the *Darlington Conference 1998*.

3. Jennifer Frautschi, Private communication.

4. *Antonio Stradivari; Master Luthier. His Life and Instruments* by W. Henley (Amati Publishing Ltd., Brighton 1. Sussex, England).

5. Fry emphasizes how important it is for young students to have a good instrument. *"I have seen some cases where students have obtained a good violin and their ability has been radically changed by having a good instrument... It is the time when they are learning that it is important to have a good instrument because they don't need to struggle with the problems of a poor violin as well as learning the technique playing it."* Laura Burns is a recent example of a young violinist who plays a Fry violin that was reconstructed with the sound of a Cremonise Seraphin in mind. Laura is a member of the Madison Symphony Orchestra, where she also performs with the Rhapsody String Quartet, part of the Madison Symphony's Heartstrings initiative that brings live interactive music programs to adults and children with disabilities. She has been playing on this violin ever since she played it during her first performance at the Token Creek Chamber Music Festival in August 2008. *"The clarity and focus in the sound was phenomenal,"* she says. *"I notice that passages are clearer when I play. I spend my practice time working with different colors of sounds, instead of trying to find a way around the problems of my instrument. It gives me great satisfaction knowing that I*

do not need to make musical compromises because the instrument is not able to produce a sound, or respond with clarity."

6. Among the great instruments of Cremona, no two instruments are alike. They have individual personalities with tonal qualities some good, some not so good. In the light of this, Fry hopes some outstanding players with open mind will judge and say, *"Your instrument has good response. It has the quality that I would put it in the category of a good Stradivari."*

Chapter 8

1. *Measure for Measure; A Musical History of Science*, Thomas Levenson (Simon & Schuster, New York 1994).
2. Conversations with Fry.
3. Martin Schleske, *Wired for Sound, The Strad October 2002*: pp. 1111–1114.
4. Fry is leaving behind over six hundred pages of his notes and his experimental records in the Rare Book Section of the University of Wisconsin Library with free access to any one interested. Also a Video Book concerning Simple Physics of Violin is in preparation.
5. A fascinating subject that touches many things as Fry says. *"It touches aesthetics, the whole subject of beauty; it touches science because we want to understand; it touches the emotion of people because it does stir up emotions when we hear a beautiful instrument played well."*

Acknowledgments

First and foremost, I am enormously grateful to William F. "Jack" Fry, for innumerable hours of extended conversations concerning his research spanning several decades, providing me with all the necessary drawings of figures and also for his participation in the video. It has been a very enjoyable and endearing collaboration. My thanks to Audrey Fry for her account regarding Fry's early life during his school days.

Special thanks to Rosemary Harbison, Anne Houssay and Aleth Michel. To Rosemary Harbison, an accomplished concert violinist, for many helpful discussions concerning her vital role in Fry's research and her participation in the video accompanying the book. Anne Houssay, "conservatrice" at Cité de la Musique in Paris, helped me immensely with background research on violins, securing access to the facilities at the Cité museum and documentation center. Likewise, I am indebted to Aleth Michel, who, through her trial and tribulations in becoming a noted violin maker, has provided me deep insights into the art of violin making; additionally, for making the literature and conference proceedings of violin makers available to me.

My thanks are also due to Janet Pease, Shrila Pradhan and Mary Waitrovich — Janet, for her transcription of hundreds of hours of taped conversations, their edits and her help in research, Shrila, my editor, for her excellent job and her enthusiastic support and Mary Waitrovich, the Digital Media Co-ordinator at the University of Wisconsin for the making of the video. I have been fortunate to have the help and advice of Tess Gallagher for her critical reading of parts of the manuscript. Thanks also to Monona Wali, whose cumulative reading of the manuscript added the needed finishing touches.

Over the years, I have had informal, very helpful conversations with several violinists concerning Cremona violins and the instruments they played. My special thanks to Joshua Bell, Jennifer Frautschi and Midori. I also interviewed several violin makers and violin researchers — without their generous sharing of time and enthusiastic support, this book would not have the character it has. I list their names alphabetically below. My thanks are also extended to friends, Lynn Margulis, Laura Fenster, Laurie Brown, Gary Goldstein and Dzidra Knecht, who have been a constant source of support and encouragement.

Finally, my thanks to Heather Kirkpatrick for preparing the manuscript, getting all the figures and pictures ready and Senior Editor Lakshmi Narayanan and her colleagues at World Scientific Publishing for the excellent job they have done in producing this book.

Interviews

Name	Profession	Date Interviewed	Place Interviewed
Rachel Barton	Violinist	May 8, 1994	Chicago, IL
Graham Caldersmith	Violin maker	April 28, 1995	Canberra, Australia
John Ferwerda	Violin maker	April 1, 1995	Melbourne, Australia
Norman Fletcher	Violin research	April 29, 1995	Canberra, Australia
Cyrus Forough	Violinist	October 22, 1995	Northbrook, MA
Carleen M. Hutchins	Violin research	November 24, 1996	
Rosemary Harbison	Violinist	February 7, 1991	Cambridge, MA
		June 19&20, 1993	
Martin Andrews	Violin Workshop	March 1990	Bristol, England
		March 23, 1996	
Aleth Michel	Violin maker	March 1990	Bristol, England
		March 23, 1996	
Kim Won-Mo	Violinist	June 17, 1993	New York, NY
Stephen Novak	Violin maker	March 1990	Bristol, England
Gabriel Weinreich	Violin research	December 7, 1996	Ann Arbor, MI

Selected Bibliography

Boyden David, *The History of Violin Playing. From its origins to 1761 and its Relationship to the Violin and Violin Music*, Clarendon Press, Oxford, Oxford University Press, New York, 1990.

Coates Kevin, *Geometry, Proportions and the Art of Lutherie*, Clarendon Press, Oxford, 1985.

Cremer Lothar, *The Physics of the Violin*, The MIT Press, Cambridge, Massachusetts, 1984.

Faber Toby, *Stradivari's Genius*, Random House, New York, 2004.

Farga Franz, *Violins & Violinists (Translated by Egon Larsen)*, ROCKLIFF, Salisbury Square, London, 1940.

Harvey Brian W., *Violin Fraud, Deception, Forgery, Theft, and the Law*, Clarendon Press, Oxford, 1992.

Helmholtz Hermann, *On the Sensations of Tone*, Dover Publications, Inc. New York, 1954.

Henley William (*Revised and edited by* C. Woodcock), *Antonio Stradivari*, Master Luthier; CREMONA ITALY 1644–1737, *His Life and Times*, Amati Publishing Ltd. Brighton 1. Sussex. England, 1961.

Heron-Allen Edward, *Violin-Making, As it Was and Is*, Ward Lock & Co., Limited, London, Melbourne and Cape Town, 1885.

Hill Henry W., Arthur F., and Alfred E., *Antonio Stradivari, His Life and Work (1644–1737)*, Dover Publications, Inc. New York 1963, *The Violin-Makers of the Guarneri Family (1626–1762)*, Dover Publications, Inc. New York, 1989.

Holman Peter, *Four and Twenty Fiddlers*, Clarendon Press, Oxford, 1993.

Hutchins Carleen Maley Ed., *Research Papers in Violin Acoustics 1975–1993*, Published by the Acoustical Society of America through the American Institute of Physics.

Hunt Frederick Vinton, *Origins in Acoustics*, Published by the Acoustical Society of America through the American Institute of Physics, 1992.

Jalovec Karel, *Italian Violin Makers*, Anglo-Italian Publications Limited, London.

Kolneder Walter, *The Amadeus Book of the Violin*, Construction, History, and Music, Amadeus Press, Portland, Oregon, 1998.

Levenson Thomas, *Measure for Measure: A Musical History of Science*, Simon & Schuster, New York, 1994.

Reid Joseph V., *You Can Make a Stradivarius Violin*, William Lewis & Son, Lincolnwood, 1907.

Roth Henry, *Master Violinists in Performance*, Paganiniana Publications, Inc. Neptune city, NJ, 1982.

Sacconi Simone F., *The "Secrets of Stradivari"*, Libreria del Convegno, Cremona, 1972.

Silverman William Alexander, *The Violin Hunter*, The John Day Company, New York, 1957.

Schleske Martin, *Tonal Copies in the Field of Violinmaking*, Meisteratelier für Geogenbau. Grubmühl 22 D-82131 Stockdorf/München.

Stoeving Paul, *The Story of the Violin*, The Walter Scott Publishing Co., Ltd. New York: Charles Scribner's sons, 1904.

Stowell Robin, *Violin Techniques and Performance Practice in the Late Eighteen and Early Nineteenth Centuries*, Cambridge University Press, 1990.

Venkataraman G., *Journey into Light, Life and Science of C.V. Raman*, Penguin Books India, 1994.

Wechsberg Joseph, *The Glory of Violin*, The Viking Press, New York, 1972.

Wibberley Leonard, *Guarneri, Violin-maker of Genius*, Macdonald and Jane's London, 1974.

Index

About the Author

Kameshwar C. Wali is the Distinguished Research Professor Emeritus in the Physics Department, Syracuse University, Syracuse, NY. He is a theoretical physicist specializing in high energy physics. Born in India, he has been in the United States since 1955 and first met William F. Fry at the University of Wisconsin, Madison, in 1955. He is the author of *CHANDRA: A biography of S. Chandrasekhar*.